CAMBRIDGE LIBRARY COLLECTION

Books of enduring scholarly value

Earth Sciences

In the nineteenth century, geology emerged as a distinct academic discipline. It pointed the way towards the theory of evolution, as scientists including Gideon Mantell, Adam Sedgwick, Charles Lyell and Roderick Murchison began to use the evidence of minerals, rock formations and fossils to demonstrate that the earth was older by millions of years than the conventional, Bible-based wisdom had supposed. They argued convincingly that the climate, flora and fauna of the distant past could be deduced from geological evidence. Volcanic activity, the formation of mountains, and the action of glaciers and rivers, tides and ocean currents also became better understood. This series includes landmark publications by pioneers of the modern earth sciences, who advanced the scientific understanding of our planet and the processes by which it is constantly re-shaped.

Seven Lectures on Meteorology

Luke Howard (1772–1864) was a pharmacist and businessman, but is most famous for his contributions to meteorology. He classified clouds by their appearance and gave them their modern names of cumulus, cirrus, nimbus and stratus. He was educated at a Quaker school in Oxfordshire, then trained as a pharmacist, but was fascinated by weather throughout his life, and developed into a keen amateur meteorologist. He wrote several important texts on the subject including *The Climate of London*, an early study in urban climatology, and *On the Modification of Clouds* (both also reissued in this series). Published in 1837, *Seven Lectures on Meteorology* covers the components of the atmosphere, seasonal variation in winds and temperature, the use of barometers, cloud structure, and visual phenomena such as rainbows and the Aurora Borealis. This reissue also includes Howard's short 1842 book which details selected British weather data from 1824 to 1841.

Cambridge University Press has long been a pioneer in the reissuing of out-of-print titles from its own backlist, producing digital reprints of books that are still sought after by scholars and students but could not be reprinted economically using traditional technology. The Cambridge Library Collection extends this activity to a wider range of books which are still of importance to researchers and professionals, either for the source material they contain, or as landmarks in the history of their academic discipline.

Drawing from the world-renowned collections in the Cambridge University Library, and guided by the advice of experts in each subject area, Cambridge University Press is using state-of-the-art scanning machines in its own Printing House to capture the content of each book selected for inclusion. The files are processed to give a consistently clear, crisp image, and the books finished to the high quality standard for which the Press is recognised around the world. The latest print-on-demand technology ensures that the books will remain available indefinitely, and that orders for single or multiple copies can quickly be supplied.

The Cambridge Library Collection will bring back to life books of enduring scholarly value (including out-of-copyright works originally issued by other publishers) across a wide range of disciplines in the humanities and social sciences and in science and technology.

Seven Lectures
on Meteorology

LUKE HOWARD

CAMBRIDGE UNIVERSITY PRESS

Cambridge, New York, Melbourne, Madrid, Cape Town,
Singapore, São Paolo, Delhi, Tokyo, Mexico City

Published in the United States of America by Cambridge University Press, New York

www.cambridge.org
Information on this title: www.cambridge.org/9781108040167

This edition first published 1837
This digitally printed version 2011

ISBN 978-1-108-04016-7 Paperback

SEVEN LECTURES

ON

METEOROLOGY,

BY

LUKE HOWARD, Gent., F.R.S., &c.

AUTHOR OF AN ' ESSAY ON THE MODIFICATIONS OF CLOUDS,'
' THE CLIMATE OF LONDON,' AND OTHER WORKS
ON THE SUBJECT.

Actorum sic juvat meminisse laborum.

PONTEFRACT:

PRINTED BY JAMES LUCAS, MARKET-PLACE.

1837.

J. LUCAS, PRINTER, MARKET-PLACE, PONTEFRACT.

TO

JOHN DALTON, Esq., D.C.L., F.R.S.,

&c. &c. &c.

IN TOKEN OF A FORTY YEARS' FRIENDSHIP BETWEEN US,

ORIGINATING IN THE

CULTIVATION BY BOTH OF THE SCIENCE HERE TREATED,

THIS LITTLE VOLUME IS INSCRIBED

BY THE AUTHOR.

PREFACE.

IT is now twenty years since the author, having delivered a course of Lectures on Meteorology to a select party at Tottenham, gave expectation of making them public : the course has been repeated on different occasions, and to like companies, since. The author's design in forming, from his notes and with the help of previous publications, the present work, is to present in a familiar and intelligible way the facts and principles of this art and study ; now becoming also a science. He does not pretend, however, to satisfy the requisitions of SCIENCE by his manner of treating the subject: though teachers of a higher grade may possibly find in his work a foundation capable of such a superstructure. He has no further ambition of his own, than to instruct and entertain those, who may be prepared for the course by an ordinary degree of acquaintance with the principles of natural philosophy, and the terms of his art.

It has been judged convenient to make the text *continuous*—that the Lecturer may more readily find it again after breaking off to demonstrate: and to put what is for *his* particular information, or guidance, in a set of notes *at the end*.

iv

Here will be found also portions of matter, and references to more, which may serve on occasion to diversify a Lecture, by being substituted for a part of the text. In making his *references*, the author has confined himself very much to two sources: the collection of facts and observations which he had previously embodied in his *Climate of London;* and the *Philosophical Transactions:* the former portable, in 3 vols. 8vo., 2nd Edit., 1833—the latter accessible in every public scientific Library, or Institution. It would have been possible, with time and pains, to have sent the reader into a much wider field—but it is doubtful whether to advantage, *for the purpose in hand.*

Such as the work is, (prepared and got up in a remote country place, under many disadvantages of execution,) the author submits it to his friends and associates in Science; inviting, and quite disposed to receive in good part, whatsoever observations or hints for its improvement in a manner consistent with the original design, they may be pleased to communicate: and of which, should his life be prolonged and ability left him, he may avail himself for that end in another Edition.

The Villa, Ackworth,
17th Fourth Month, 1837.

LECTURE FIRST.

Constitution and properties of the Atmosphere.

The subject of the Course of Lectures I am now about to commence is METEOROLOGY, or the study of the air, weather and seasons, with the varied and interesting appearances which they present. To persons unfurnished with the principles of this kind of knowledge, it is too common to regard it with indifference; or to bestow on it no more than the supercilious notice they would take of the Weather-column in a Moore's Almanac. As if there were not in our nature higher motives to the inquiry into these things, than our convenience during a walk or a journey; or even than the interest we feel in the success of our agricultural pursuits, or of those adventures which the hardy mariner, or the fisherman, are daily prosecuting in our behalf. To these persons, the necessity of their condition prescribes the observance of the face of the sky, and of the changes of the fluctuating deep, *as a part of their education:* they become weather-wise by tradition and experience; and are often able to communicate the results of a certain local knowledge, without being prepared to assign *a reason* for any thing they say.

But the rest of CREATION is studied in order to be *known:* and many ingenious persons engage in pursuits connected with Natural History and Philosophy, the immediate utility of which they would be very much at a loss to demonstrate. The starry heavens are yet scrutinized with minute exactness, and the returns of their periodical phenomena registered anew; though Navigation, as an art, may be said to be perfected; and though it be many ages since we became able to measure time backward and forward to any extent, and to predict with certainty the occurrence of Eclipses, and the return of Comets to our system. The face of every region of the globe is to be found presented in the descriptive accounts of Travellers, and mapped to shew its extent and boundaries, its cities, mountains and rivers. Of later time, also, the Geologist, following the miner beneath the surface, or observing its upward throws, its wastings and subsidences, has enabled us to contemplate the internal structure of the earth, and to see as it were the very foundations of the hills we tread on.

The *sky* too belongs to the Landscape;—the ocean of air in which we live and move, with its continents and islands of cloud, its tides and currents of constant and variable winds, is a component part of the great globe: and those regions in which the bolt of heaven is forged, and the fructifying rain condensed, —where the cold hail concretes in the summer cloud,—and from whence large masses of stone and metal have descended at times upon the earth,—can never be to the zealous Naturalist a subject of tame and unfeeling contemplation.

If we rise higher, and mix devotion with the feeling of rapture which the grandeur and loveliness of

Nature is so apt to inspire, we shall find in Holy
Scripture " the balancings of the clouds " mentioned
as a part of the wondrous works of Him who is
perfect in knowledge. The nice adjustments of cause
and effect are here proposed to us, as proofs of
Creative energy and skill; as an argument that God
himself intends that we should minutely examine,
and critically prove the perfection of his handy work,
(as Himself was pleased to do at its formation), and
deduce from the search further matter of admiration
and praise. Were it not so, He would not have
bestowed upon us that curiosity to observe and
know, that sagacity in making inferences, that comprehensive faculty of combining results and forming
systems, which exercised in all ages by the few for
the benefit of the many, has at length spread before
mankind " his book of knowledge fair,"—so widely
spread it, that in vain would the most powerful intel-
lect strive to grasp the whole : and they who would
learn perfectly must choose each one his page.

A diligent, though not a profound student, one who
is satisfied to avail himself of the discoveries of
those who have gone before him, may thus present
in simple arrangement, and enable others to com-
prehend with ease, things the Natural history of which
it has cost Science a world of pains to analyze and
recompose : and such an exposition of established
facts is, precisely, that which he who undertakes the
work of public instruction has to perform. I may
be permitted here to remark, for the encouragement
of students, that by applying for a time exclusively
to some simple department of knowledge, and making
it the subject of a regular course of inquiry, the man
whose daily occupations afford him but little leisure,
may yet find that little fruitful of improvement.

Persevering labour, it is said (even unaided by experience) overcomes all difficulties: *(a)* and without
toil and pains who would expect in any difficult
enterprise to succeed ? Moreover, the more dry and
uninviting the subject in its opening, the more
strongly will the student find himself attached to it
when, having conquered the first difficulties, he is
sensible of being in possession of some portion of
demonstrable truth.—But we must now come to our
matter.

The air we breathe, is a transparent, elastic-fluid
substance, covering uniformly this globe of land and
water to the depth, as is supposed, of about fifty
miles : (*b*)—more at the Equator and less at the poles.
In its collective form, it is denominated THE
ATMOSPHERE ; a term composed of two Greek words,
signifying *a surrounding vapour*. It is borne along
with the Earth about the Sun ; and, along with the
surface on which it rests, about the Earth's axis.

If we take the diameter of our Earth at eight
thousand miles, and the depth of the body of air
resting on it at the Equator, (where it is most ex-
panded), at eighty miles, the proportion of the
atmosphere to the *sphere* will appear in section as
in this Figure. *(c)* This apparently thin and scanty
covering (sufficient however for all the purposes of its
formation) presses on the surface with a weight
equal at the level of the sea, to fifteen pounds on
every square inch ; and which amounts in the aggregate to twelve millions of millions of pounds, or
to that of a solid globe of lead of sixty miles in
diameter.

Small portions of the air may be weighed in a
balance, *(d)* and its weight may be otherwise proved

(a) For the *Notes*, see the Appendix at the end.

hydrostatically, by shewing how it bears up solids immersed in it, as water does. *(e)* That it is a *substance,* we need not attempt to demonstrate to any one who has felt the force of a strong wind. Nothing unsubstantial could propel ships on their course, or carry about their axis the heavy sails of a windmill. That it is a *fluid* substance, we may prove to ourselves with a pair of *common bellows.* In the bottom board of these is a valve of large diameter, which admits air quickly as I lift the top one. When I press this down again, the valve shuts and the air is expelled, through the smaller opening at the nozzle, exactly as water from an engine; and in point of *force* with a similar effect. *(f)*

The substantial quality of air is however more commonly shewn by its *resistance;* and we have an elegant method of demonstrating this by experiment. Here is a little machine, containing two sets of brass paddles, set on as in a mill-wheel: the one with its flat sides, the other with its edges opposed to the air (when set in motion), and in all other respects alike. I shall first give them both an equal impulse in the open air, by means of the rack and pinion attached to each. *(g)* You see that the wheel with the flat sides opposed stops in about half the time that the other does. I shall now expose them to the like equal impulse in a *vacuum;* that is, in a space from which I shall have first taken out the air by an air-pump, till only about a sixtieth part remains:— which is as much as we can conveniently do on this occasion, and sufficient for our purpose. You see that, now, the two stop very nearly at the same moment. The reason, then, of the retardation, and final earlier stopping, of the wheel having the flat sides opposed to the air, was the resistance opposed

by the latter *as a fluid medium :* and *water* would,
of course, occasion both wheels to stop in less time
than air, (*h*) as affording a greater resistance. A
Cannon ball, fired with the ordinary charge of powder, has to overcome in its course through the air,
from this cause, a resistance exceeding twenty three
times its own weight.

As a fluid *penetrable in all directions,* like the
ocean on which it rests, the air is the great highway
of the feathered and insect tribes.   It is dense
enough to bear them up to great heights, (*i*) and yet
so thin as to afford but little impediment to their
movements. (*k*)

As a fluid, again, it is *the ordinary vehicle of
sound,*—not the only or exclusive one; for *water*
is known to convey sound very well, not only to its
proper inhabitants the fishes, but also to the human
ear. (*l*)   Lastly, as to the further properties that
concern our present subject, the air is the ordinary
(but again not the exclusive) means and support of
combustion ; so that, without it, we could have
neither convenient *artificial light,* nor *fire* to afford
us heat. (*m*) And it is (as we sensibly feel in a close
room) the necessary means of *respiration :* the experiments to prove this, to an extreme point, involve
too much of suffering to the animals made the subject
of them, to be encouraged by a public exhibition. (*n*)

Before proceeding further, we may as well demonstrate experimentally the *pressure* of the atmosphere.
Here is a piece of thin membrane (*o*) tied over the
mouth of a glass made open at both ends. I shall place
the other end on the plate of the air-pump, and exhaust the air from beneath it. You see that the
membrane is first depressed by the weight of the
incumbent air, (which may amount to some forty or
fifty pounds) and, then, *broken.*

As fluids press alike in all directions, according to the incumbent weight, we shall be able by the same means *to break a glass bottle.* The air contained in this square bottle (which is made thin for the purpose) will now be exhausted, along with that contained in the glass under which I place it. (*p*) When I restore the pressure by re-admitting the air, it being prevented by a *valve* from passing also into the *bottle*, the latter (if not made too strong, as they sometimes are,) will be broken.

In consequence of this property, the air is found in situations where one would least expect it   The tenuity of this fluid, and the great force applied by the pressure, make it penetrate the substance of both the solids and liquids of which the earth is composed, to an unknown depth.   I shall take off the pressure of the atmosphere from a glass of spring water, newly drawn from the pump, at the bottom of which is a lump of marble.   You observe that the air is first given out from the water ; which is rendered turbid by the minute bubbles, and at length appears to simmer at the top.   Now, the water having cleared itself, the *marble* shews bubbles of air, and detaches them from its surface : these would continue to be emitted for a long time, should we keep up the *vacuum ;* and were we perfectly to clear the pores of the marble, and readmit the air into the glass, the *water* would be forced into the substance of the marble in place of the air, and would give it the transparent look of alabaster.   Spars and cristals, formed out of a watery medium, contain a portion of *water of cristallization* in their proper substance, making them always transparent : but dry stony concretes contain air in their pores ; and are hence opake.   This same water, set in a cool cellar, would

be *aerated* again in the space of twenty four hours,
by the mere pressure of the incumbent atmosphere ;
and much more speedily, if it were shaken.

Growing trees and other vegetables, with their
products, contain air in abundance: and fruits
which have shrivelled (if the skin be sound) may be
plumped again, by taking off the external pressure
from the air within them.(*q*) Even oils, when obtained
from vegetable matters, such as *linseed*, by pressure
without heat, are found by this test to abound with
air. (*r*)

The most useful *sensible* effect of this pressure
that we experience is, the raising of water in the
barrel of the common pump.   Here is a glass model
of a pump, fitted with the usual valves.; one of which
is placed in the bottom of the pump, the other in
the piston.   I shall place the model on the top of a
glass jar ; having first screwed on this pipe, which
descends to the bottom of a glass of water, standing
within the jar and representing the well : the pump
has been previously filled with water. When I move
the handle *up*, the valve in the descending bucket
(or piston) opens, and lets the water from beneath it
through : when I press down the handle, this valve
closes, the lower one opens ; and first the air, and
then the water from the glass rush through it ; and
as the lower valve closes again, upon the descending
stroke of the piston, pass through to be discharged
at the spout.   The pressure of the air (to which I
have taken care, hitherto, to afford a passage to the
surface of the well) is the cause of the water's rising,
to fill the vacuum which ensues upon the upward
movement of the piston.   This I shall immediately
demonstrate: I have now secured the jar, and created
a sort of vacuum around the water, by working the

air-pump: you now see the piston moving up and down without effect, the water passing by its sides:— or, if these be tight, or the movement too quick, there is a sensible recoil upon the vacuum in the pump. I shall now readmit the air, and you will find that the water from our little well rises again freely. (s)

Unlike water, the air is *elastic:* and this is by far the most remarkable of its properties, as an element of nature. To shew what is here meant by the term *elasticity,* (t) I must have recourse to experiment. Here is a flaccid bladder, containing a little air, well secured, and placed under weights in a frame. When I remove the air from about it by the pump, you observe that the bladder begins to swell, by the expansion of the contained air; and at length lifts a considerable weight, to the extent permitted by the frame. (u) Again, I shall place on the plate of the pump this bottle, containing air with a little quicksilver under it, into which dips an open glass tube, the neck of the bottle being well closed about the tube. I have covered it with a close jar and tube, and exhausted a portion of the air from the containing vessel: the air in the *bottle* is unable to follow that portion, yet, finding elbow-room, you see it expands, and raises the quicksilver in a column in the inner tube: we shall have occasion to repeat and enlarge upon this elegant experiment, with a different object, in a future lecture. (v)

The expansion of the air is due, in each case, to a property inherent in its particles of repelling each other: and it may be further illustrated by the action of a watch-spring; which opens itself out the moment it is set at liberty from the bounds that con-

fined it. To this property we owe the facility of
exhausting, or pumping out, the air from the Receiver of the air-pump, by means of a detached
barrel or barrels, which are fitted with valves very
much in the manner of a water-pump. (*w*) The air
expanding, as successive portions are driven out at
each stroke of the piston, and entering continually
the vacuum created in the barrel, may at length, with
a very perfect machine, be thus reduced to a three-
hundredth part, or less, of its original quantity and
pressure: yet retaining its *bulk*, and filling the
Receiver as at first. On the like principle we might,
by reversing the valves and somewhat altering the
construction, introduce from without, and *condense*
in the receiver, many times the volume of air which
it would naturally contain: the elastic property of
the outer air is, here again, the cause of its passing
the valve, *into the barrel of the pump*, as oft as a
vacuum is there made, to be thrown thence into the
Receiver.

Were not this expansive force counteracted by the
inherent *gravity* of the particles, we should lose the
greater part of our Atmosphere, in a single circuit of
the Earth about the Sun: all would be dissipated
into surrounding space. But the gravity of the solid
tho' invisible particles of elastic fluids, serves to keep
them in their proper relation to the mass of our
globe: and the manner of this arrangement is what
we have next to explain. I have spoken of a steel
spring: let us try a sufficiently delicate specimen
of coiled wire, laid on a table on its side—the
coils will be found to keep their relative distances
equably, through the whole length of the coil: but
set the whole on end, and the effect is altered—the
weight of the upper coils now confines the elastic

power of the lower, in such a way as that they approach nearer and nearer together, as we descend through the series to the bottom of the coil. Thus it is with the air:—it is rarefied (as we express the effect) in proportion as the weight of the superior air becomes less. On a mountain, it is hence found to be much thinner than on the plain. If we carry air from the latter, confined in a bottle, to the mountain top, and there open the bottle, it will rush out with some force: if we now close the bottle and descend with it, and open it with the mouth dipped in water, the water will enter, and fill a portion of the space voided by the air.

The rarefaction is, however, not in direct proportion to the heights, but much greater. If at seven miles height, air would occupy four times the space it did below, at fourteen miles it would occupy sixteen times the space, and at twenty one miles, sixty four times,—and so on. Thus at fifty miles it may, by possibility, be sixteen thousand times more rare than below; and yet preserve its relation to the earth; and—the vast extent considered—a definite surface. But it is not probable that it becomes quite so rare:—there is a counteracting cause in the *cold* of the higher regions; which is found greater, in proportion as we ascend higher into the atmosphere. All elastic fluids expand by heat and contract by cold: you will see the bead of quicksilver in this tube rise rapidly, when I put the bottle into warm water; and descend as fast when I plunge it into cold water. (x) This motion is the measure of the dilatation and contraction of the air confined in the bottle, in which the tube stands: an instrument so constructed is called a *manometer*, and it makes a sort of imperfect Thermometer, shewing the natural variations of heat and pressure in the air.

I have said that the air was the ordinary vehicle of *sound :* which cannot be propagated from the sounding body to the ear without a *medium :* a very coarse one, to be sure, may seem to serve the purpose—for the scratch of a pin, at one end of a long piece of *timber*, may be heard by an ear placed at the other:—but it is done still with the help of the air adhering to the surface of the wood. Here is a small bell, having a clapper connected with a toothed wheel and pallets : when I turn the wheel, by means of a string passed round a pulley, and ring the bell under this glass, you hear the sound very distinctly : I will now exhaust the air to about half an inch of pressure, and you will not be able to hear the bell when struck by the clapper, at all. (*y*)

Hitherto we have regarded the vast äerial ocean as at rest : we have now to view it in motion, and to contemplate the means by which its internal currents are kept up, and its parts mixed or exchanged, by a constant circulation between the Equatorial regions and those towards either Pole. All our principal *Winds*, (so denominated from the winding motion of air) originate in the expansion caused by the Solar heat : partial movements and gyrations may be produced by the evaporation and condensation of water, by Electrical attractions, and by other natural causes—of which hereafter : but these currents do not affect large portions of the atmosphere.

We will avail ourselves of an artificial cause, the heat of a common fire, (or of our bodies,) to shew the manner in which the atmosphere may be affected on the great scale by heat. The air in the adjoining room (or passage, or street) is, I presume, colder than that in which we are assembled. I shall set open the door, and hold a lighted candle in the bottom

of the door-way. The air, you observe, here passes in a
pretty strong current inward : and when I remove the
candle to the *top* of the opening, the flame is carried
outward, indicating the passage of a current in that
direction. One of these is the colder and *denser* air,
displacing that of the room by its greater gravity—
the other is the warm air flowing out, *over the cold:*
and it will not be difficult to find a station, about
midway between the two, in which the flame will
burn upright, that stratum of the air being at rest.
In this way it is, that apartments are ventilated by
the door alone being set open : the contained atmosphere being speedily exchanged by means of these
opposite currents for a body of cooler air from without.

You will be ready perhaps on the mention of
ventilation to advert to the *fireplace,* and to the
draught of air towards it; which is also, and more
commonly, the means of keeping the air fresh in a
room : but the *principle* is still the same. The air
is not *drawn* up the chimney, by any power seated
there : it is forced up, by the *pressure* of the cold air
from without on the warmer and less dense air of
the room. And it *may* happen, if the house be too
well closed, that the most convenient passage for the
supply of air may be *down* some neighbouring cold
chimney, to go up the heated one. In this case,
vain are all endeavours to make a fire *draw* in the
second chimney, until an opening has been provided
for the entrance of the outer cold air in another way.
Count Rumford taught us, about the beginning of the
present Century, how to secure the draught of our
fireplaces, by excluding the unnecessary passage
into the chimney of superfluous cold air from the
room : and to the general attention originally paid,
by ingenious workmen, to *his* directions, we are in-

c

debted for a better construction of these conveniences:
which has generally had the effect of removing from
our houses that proverbial domestic nuisance, a
smoky chimney!

Thus it is that our *great currents* also are set
going—and *kept* going: for, that the bulk of the
atmosphere is in continual motion, is matter of common experience. Ships sail not merely to distant
parts but to the antipodes, *and return in a given
time:* they could not do this, if brisk winds were
not in general to be relied on, every where, at sea.
Calms of large extent or of long duration, great stagnant pools as it were in the äerial ocean, are not
commonly met with, and are more dreaded by sailors
than a storm. The invention of Navigation by *steam*
may indeed serve to make many of us less observant
of this fact: but it is still notorious, even to lands-
men, how constantly it blows from *some* point; and
how our coasters are detained at times, (and for a
long time together at particular passages) *by adverse
winds.* (z)

I shall reserve for the next Lecture the mention
of the particular winds of different latitudes and
climates; it may suffice here to describe the general
effect of the Solar heat upon our atmosphere. The
air, then, in the temperate zones on each side the
Equator, being constantly cooler, and hence *denser* or
heavier than the air in the Torrid zone, presses upon
the latter from North and South, in currents that
flow in the lower atmosphere, and are always to be
met with *as winds* in certain latitudes at sea. The
Tropical air, heated and rarefied, is continually
driven upward, and returns (as the phenomena of
the winds in Temperate Latitudes demonstrate) *over*
these converging colder currents, back to the South

and North; becoming sensible, *as a wind,* only when it again reaches the surface of the land or sea in its descent. By the various circumstances of the seasons, and of planetary attraction, these vast streams, constantly going and returning in the atmosphere, are brought at times more over the sea, at others more over the land ; and are extended and contracted in their course :—now impinging with tempestuous violence on our own shores, now breaking up and melting the polar ices: and again clearing and cooling our atmosphere by the importation of an air which but a little before was freezing within the Arctic Circle. For we must not forget that *cold* here operates along with the resistance afforded by the surface, in lulling the tempest, and casting the air into a dead sleep, from which it is roused again by the approach of the Solar rays in spring. This fact is so well known to mariners who frequent the Polar ocean, that they have only to moor their ship *in the lee* of a large pack of ice, to be quite sheltered from a furious gale blowing from the Southward on the other side of it. Perhaps the differing *electricities* of the air and ice may tend, in these cases, to bring them into more effectual contact ; but it is too early in our course to indulge in such speculations.

The consequence however, and a most important and salutary consequence, of this arrangement is, that no region of the globe can long retain its own atmosphere ; but is obliged to part with it by an ascending or a lateral movement, and receive that of some distant region in exchange. And the benefits hence resulting to all countries in the facility of intercourse afforded them ;—in the alternate production of rain and drying of the earth's surface—in the prevention of a noisome, putrid pestilential state of

atmosphere, in the invigoration and comfort of our
animal frames by the constant change,—are too
obvious to need a particular enumeration, and too
sensible as the allwise Creator's work not to claim
our admiration and praise.

The ancients had to contend with a great difficulty
in addition to imperfect science, *in their super-
stitious fears.* Knowing but little of natural causes,
they were timid in proportion to their ignorance,
and scarce dared to venture from the coast, although
there the *dangers of Navigation* are chiefly found.
For example as to their superstitions,—they had a
story of one Æolus, son of Hippotas, king of those
now called the Lipari isles (or Æolia,) who kept
the winds in a *cave* in a huge *mountain,* and loosed
them at his pleasure; to afford a passage to the
mariner, or to ruin him by a storm. This *god of the
winds* is made by Homer to give Ulysses the winds
he wanted for his return home from the wars of Troy:
they were tied up, it seems, in bags, to be loosed for
use at particular points of the voyage—but his men
chancing from curiosity to open a bag prematurely,
he gets shipwrecked in consequence. Æneas, again,
the hero of Virgil, having escaped from Troy and
being on his voyage from Sicily towards Latium, the
poet makes Æolus, at the request of the queen of
heaven (the enemy of Troy) strike his sceptre into the
mountain side, making a huge opening, through
which rushed in all their fury, Eurus and Africus
and Notus (ruffian blasts from every point of the
compass) to sink and disperse his fleet upon the
Tyrrhene sea !

" Talia flammato secum Dea corde volutans,
Nimborum in patriam, loca freta furentibus Austris,
Æoliam venit. Hic, vasto rex Æolus antro
Luctantes ventos, tempestatesque sonoras
Imperio premit, ac vinclis, et carcere frenat.

Illi indignantes magno cum murmure montis
Circum claustra fremunt. Celsâ sedet Æolus arce
Sceptra tenens, mollitque animos, et temperat iras.
Nî faciat, maria ac terras cœlumque profundum
Quippe ferant rapidi secum, verrantque per auras.
Sed Pater omnipotens speluncis abdidit atris,
Hoe metuens: molemque et montes insuper altos
Imposuit, regemque dedit, qui fœdere certo
Et premere, et laxas sciret dare jussus habenas.

Æneidos i. 50.

We have, I think, instances in more modern accounts of voyages and travels, of wizards on the
coasts of the North of Europe, who would *sell* a
wind to the credulous mariner; tying certain knots
upon his rigging, to be loosed at certain points or
promontories at which he should arrive:—a result,
probably, of a thorough local knowledge of the bearings and trendings of a dangerous coast, and of the
effects of every headland upon the general wind prevailing at the time: as this poetical flight of Virgil's
might have reference to the dangers of doubling
Cape Spartivento too near the land.

Let me now conclude this First Lecture by calling
your attention to the many *use*s of this äerial covering of our globe. Supposing mankind and other
animals, and even vegetables, to have been so formed
as not to need it for respiration, for nutrition and for
warmth, still, *without the air* what a dull scene,
what a blank in nature, in place of our many
enjoyments abroad! No refreshing breezes, no blue
skies alternating with kindly showers, no waving
branches and rustling leaves: none of the beauty
and variety of summer clouds, no rainbow, no rain!
Further, we should have lost the flight of birds with
all their cheering music, the hum of insects, the
journies of the laborious bee and the taste of honey
from the hive:—we must have conversed by signs as

deaf people ; must have dwelt by the streams, for we
could not have driven a wind-mill; and have journied
by land, for we could not have sailed so much as a
boat.   In short, the delights of sense, and the com-
merce of nations dwelling in different and distant
countries; the perfection of science and arts, and
the diffusion of that greatest of all blessings true
Religion, would have been greatly impeded and lessened: and the life of man by far less furnished with
the means of lawful and reasonable enjoyment.   Let
us then thank God that he has provided this admirable
and most beneficial part of the Creation for our use !

LECTURE SECOND.

Constant and variable Winds; Climates and Seasons.

~❦~

In our First Lecture we treated the subject of the Atmosphere at large; shewing its weight and resistance; its pressure, counteracted by elasticity; its expansion by heat and contraction by cold; its movements towards the Equator, in currents proceeding on the earth's surface; compensated by returning currents in the higher atmosphere. We shewed, likewise, that these movements originate in the rarefaction of the air between the Tropics by the Solar heat.

That the Sun's radiation is the great source of our *heat* and *light*, both, is matter of daily experience. We find at sunset, when the Earth begins to interrupt the rays of the great luminary, that both light and heat depart with him; and are restored by his return in the morning. And the mere interposition for an hour of the body of our attendant planet, the Moon in a Solar Eclipse, shall bring on a chilliness in the air (indicated also by the sinking of the Thermometer), a closing of the blossoms and leaves of such plants as open by day, and a tendency in the animal creation to silence and rest, which belong properly to the evening in its regular approaches. (*a*)

The manner and proportions of *the distribution of the Solar rays upon the Earth* form the ground of those differences of CLIMATE and SEASONS, which will constitute the principal subject of this Lecture: but before we proceed to these, we must despatch what remains of our account of the *Winds;* and first, as to their *velocity* and *force.*

A stream of air, which seamen would call a light breeze, flows at the rate of four or five miles an hour. Hence, in walking *with the wind,* we loose its cooling influence, and put ourselves as it were in a calm: on the contrary, in walking against it, we double its effect. And young men are apt to boast of outriding the breeze, and raising a wind in the opposite direction :—a thing very commonly done now, in a *pretty strong* breeze, by our locomotive engines, and the long trains attached to them.

A *fresh gale* may go at the rate of from fifteen to twenty, and a *storm of wind,* at from forty to sixty miles an hour: a good sailing boat will take for its rate, about a third of the velocity of the wind: but *ships* are not quite safe in such circumstances, and they forbear to expose the full quantity of canvas.

With respect to *hurricanes,* or such tempests in our own Latitudes as uproot trees and level buildings, and carry sheets of lead like paper through the air—it is difficult to assign them a rate; but the *direct* motion of these cannot be less than from an hundred to a hundred and twenty miles an hour. Within the *destructive whirl* the velocity is much greater; and the *force* accordingly resistless: but the account of whirlwinds belongs more properly to another part of our subject. The following are the proportions in pounds and decimal parts, of the *estimated force of pressure* on every square foot of surface, of winds of different velocities: (*b*)

Gentle breeze:	Miles, 4	.... lbs.	0·079
	5	.... —	0·123
Fresh gale,	10	.... —	0·492
	15	.... —	1·107
Strong gale,	20	.... —	1·968
	25	.... —	3·075
High wind,	30	.... —	4·429
	35	.... —	6·027
Very high wind,	40	.... —	7·873
	45	.... —	9·963
Great storm,	50	.... —	12·300
	60	.... —	17·715
Hurricane,	80	.... —	31·490
	100	.... —	49·200

Thus, when the wind blows at the rate of a mile in a minute, a wall or fence ten feet high and one hundred yards long, taken at a right angle, is exposed to a pressure equal to about sixteen and a half tons, applied to its whole surface: and the force at one hundred miles an hour might amount to nearly sixty seven tons: in which case the wall must be very strong indeed, not to be blown down. If we consider the large extent of surface opposed by well branched *trees* to the wind, (even when they are leafless), and compare it with these data, we may account for their being snapped off in the trunk, or uprooted; as they so frequently are. Another result of this so great velocity in a tempest is, that windmills if stopped in time often lose their arms; and that sometimes they cannot be stopped, but *run amain*, (as the phrase is); when the machinery *takes fire by the friction*, and they are burnt. (*c*)

Next, as to the *direction of the winds :*—it will be needful to confine this part of the subject chiefly to our own Climate, omitting all other local notices : but attending to the *great prevalent winds* of more Southern Latitudes. I have stated the general fact, of the constant approach of the air from either hemisphere towards the Equator; to displace the Tropical atmosphere, which rises by rarefaction and returns towards the poles. But the currents from North and South do not *sensibly* proceed from those opposite points, to mix under the Line. The winds, for about thirty degrees on each side, blow from North East, *on the North side of the Line ;* and from the South East, *on the South side :* and they are so constant in their course, as to be depended on by sailors, and called *Trade-winds,* from their convenience in this respect. While to the North or to the South of these limits (or outside " the Trades ") are found the *variable winds* of Temperate latitudes; partaking sometimes of the *going,* sometimes of the *returning* stream. Again, in a space of about four degrees to the North of the Line, Navigators have found the *mixing place* of the air brought from each side of it; a part of the Ocean peculiarly subject (as it seems, on that account) to *rainy calms* and *Tornadoes :*—to such changes in effect, but on a scale of much greater extent and violence, as attend our own seasons of showers and thunder: this region is called by seamen the *swamp.* The reason why the air mixes on the *North* side the Equator, appears to be this— that the Southern hemisphere is, from Astronomical causes, colder than our own : so that its air is dense to a greater extent Northward, and is hence the readiest to pour into the middle space. (*d*)

I have said that the Atmosphere is carried along with the surface on which it rests (or *moves,* for its internal movements make a *difference*, but not an exception) around the Earth's axis. You will see, while I turn the globe, how much larger this motion is upon the Equator, than about the poles. There is a point at either pole, where the air may be said to have no movement at all in diurnal revolution: as we go from either of these points towards the Equator, this movement comes on and increases. In our own latitudes, it may be six hundred miles an hour, in Iceland, only four hundred : at the Equator it must be a thousand, as the Earth is twenty four thousand miles in circumference, and revolves in twenty four hours. Now air moving, for instance, from the Arctic Circle, or from about the latitude of Iceland, to pass over the British Isles, comes continually upon ground that is, (if I may be allowed the phrase) *slipping away from under it ;*—the motion of the air in question being from North to South, and that of the Earth's surface from West to East: the consequence is, that a real Notherly stream shares an apparent Easterly direction, and becomes to us a *North East* wind. Again, let us suppose air to be transported from the coast of Guinea, or from any nearer tract *South* of us, with the rapid motion which is found in superior currents, and to descend upon the Earth's surface in these Latitudes,—such air, arriving with a share of its Westerly momentum in it could never be felt here as a true South wind; but must blow, sensibly, *from a point between South and West.*

Thus we account at once for the Trade-winds,— making only the needful changes of denomination for those in the other hemisphere ; (*e*) and we account also for the long continuance at particular seasons, of the North East and South West winds of our own Climate. We are at one time, it appears, in the tail of a steady *going* stream passing to the South; at another, are exposed to the fitful impulse of the *returning and descending current.* It should be observed, here, that while the former current, which moves next the Earth, and is kept up by rarefaction, is *a day wind*, increasing as the Sun gets up, and dying away at evening,—the latter (the South-west) has to descend from the higher atmosphere and spend its force on the earth. Every leaf that moves, every twig that bends in the gale takes away from the aggregate of its violence, (as in the experiment of the ivory balls, the ball that was at rest and is stricken *moves*, but the other that was moving is stopped in its course)—and thus is the air from the South at length quieted; and prepared by merging in a calm space, to wheel about and return. *(f)*

In the Indian ocean, and between the third and tenth degrees of South Latitude (over great tracts of sea North and South of the Line,) there are found winds called *Monsoons*, which have periodical changes; blowing during half the year *towards*, and the remainder of the year *from* the Line: these are as much depended on by sailors as the " Trades." They are explicable on the principle of *the point of greatest heat and rarefaction*, shifting with the Sun; as it approaches towards either Tropic, and becomes vertical alternately to the Northern and Southern parts of the great masses of land, seated on the middle of our globe. These winds, after a time of stormy uncertainty preceding " the change of the monsoons," take the reverse direction and follow the Sun to North or South, while circumstances are favourable to either course: the mariner being at one season in what we may call the *Polar*, and during the other in the *Equatorial* current. *(g)*

The Land and sea breeze, so commonly felt in warm climates, and sometimes observed in our own, depend in like manner on the Sun's rarefaction. The air over the land rises by day, displaced by the cooler air from the sea : at sunset, the movement is reversed—the cooler night air of the land displaces the sea air, and blows *outward* from the coast.

The third and last point for consideration respecting the Winds of our own Climate is, the proportions in which they are found to blow from different points of the Compass.   Taking them as found by the Vane, (*h*) in the neighbourhood of London, on an average of years reckoned from 1807 to 1816, these proportions in 365 days, are as follows :—

From North to North-East .... 74 days,
—— East to South-East............ 54 —
—— South to South-West........ 104 —
—— West to North-West........ 100 —
—— Various points the same day 33 —   (*i*)

This account, which is deduced from careful observation, makes the proportion of Westerly Winds to the Easterly to be as 225 to 140: and the Northerly to the Southerly, as 192 to 173—on the whole year.   It must be noted that the *direction* is here alone attended to, no reference being had to the velocity: so that the account is by no means an accurate measure of the *quantities of air passing and repassing,* over the place of observation. With regard to *Northerly and Southerly* winds, it may however be presumed that these are as nearly equal, on an average of years, as are the sums above given. But the *Westerly* preponderate so greatly, both in respect of frequency and of force, over the Easterly, that we must regard this class as containing *the predominant Winds of our Climate.*   The reason of

this may be, the great facility with which a portion of
the Equatorial or Southerly current, after descending
on the Atlantic ocean, finds its way upon the water
to our shores. Were there much land, and especially
much of wooded country in its way, we should have
less of the South-west than at present.

With regard to the average winds prevailing *in
different seasons of the year*, they may be best
given by referring them to the Months, in their order.
Northerly winds prevail most in *January*; and,
(singular as it may seem) the South-west in *February*.
In *March* and *April*, the North-east is most frequent :
but in *May*, the dominant winds become again
Southerly. *June* is distinguished by an excess of
Northerly, and *July* and *August* by a North-west
wind—which seems to be the proper *fine weather
breeze* of our summers. In *September* and *October*,
the Northerly and Southerly are balanced :—but in
November, the Northerly prevail again. In *December*, we have much more of Westerly than of
Easterly; the North and South (as in the two
Autumnal months) being in point of frequency alike.

In proceeding to the next great division of my
subject, *the differences of Climates in respect of*
HEAT *and* COLD, I shall take for granted that my
hearers are acquainted with the construction and
uses of that very common instrument *Fahrenheit's
Thermometer*. (*k*). Heat and cold are, properly
speaking, only sensations, depending on the passage
of a certain matter, or *motion* in the particles of
matter—(it is " no matter" for *our* argument which)
through the substance of our bodies. The entrance
or accumulation of this cause makes us feel the sensation of heat; the passage outward, or the deprivation
of it, that of cold. But these sensations are by no
means an accurate measure of the quantity, or even

of the *kind* of the effect produced, or of the cause present: a lump of frozen quicksilver destroys the skin as certainly as a piece of red hot charcoal, and feels very much like it in handling. But we will prove our point by a more gentle method. (*l*)

Here are three glasses, containing water at different temperatures, *or degrees of heat* : No. 1, at fifty degrees, or *cold*—No. 2, at one hundred and ten, or *hot*, (not scalding)—No. 3, at sixty three, or the warmth of the hand. I shall request some person with a delicate hand to try first No. 3, which shall be adjusted, if necessary, by adding a little warm or cold water till (with both hands or fingers in) it feels neither warm nor cold. The same person will then plunge one hand (or finger) into No. 1, the other into No. 2, and keep them immersed half a minute ; or till the sensation of heat in the one, and of cold in the other, be established. Both hands or fingers being now suddenly withdrawn, and replaced in No. 3, *the same water will be found cold to one hand and warm to the other !*

It may be asked, to what standard then do those *degrees* refer, which have been mentioned, as denoting the heat of the water in each separate glass? This is a very proper question: the reply is—it is so ordered in the economy of Creation, that *water* itself furnishes *two fixed temperatures,* one at its freezing, the other at its boiling point. By dividing the scale between these into one hundred and eighty degrees, and then extending it both ways, the Thermometers in common use in Britain, those of Fahrenheit, are made. Freezing water, or melting ice, *always indicates* thirty two on the scale: boiling water (the Barometer being at thirty inches) *always* two hundred and twelve degrees. (*m*)

Climates and seasons (our present subject) vary
from intolerable heat, down to a cold as far below
freezing as a boiling heat is above it : yet is the
feeling still comparative, depending on previous
sensations. *Campbell* found a day in Africa, with
an overcast sky and a little wind, *cool* at seventy
eight degrees—*our* full summer heat: but he had
been exposed for several days before to a heat of one
hundred and two degrees in the shade. *Parry* says,
of a Temperature *minus* fifteen Fahrenheit (or forty
seven below freezing) " it was rather pleasant to our
feelings than otherwise"—the day being fine and
calm: but he was exposed at times to a natural
cold, at which quicksilver would freeze till he could
cast a bullet of it, and fire this into a board. Thus
persons ascending a mountain ridge from the plain
are affected with chilliness, in the same inn and
room where others, who have met them in descending from the top, find the air " sultry." (*n*)

We may prove the correctness of the Thermometer
as a standard of heat, by taking equal measures (or
rather equal weights) of *hot and cold water*, at
known degrees of temperature, and mixing them: the
mixture is found to be *at the middle point of Temperature*—as for instance, water at one hundred and
twenty degrees mixed with water at sixty degrees,
will give ninety degrees by the Thermometer. We
do the same *in idea* with the air, in recording the
comparative heats of different days, or seasons, or
climates: the higher and lower observations being
added together, and divided by the number of observations taken. Thus the medium of *two* observations,
or of the maximum and minimum of heat, gives us
the mean temperature for the day:—except we choose
to be more particular, and take observations at inter-

vals through the twenty four hours. The *mean daily* observations, through the month, added up and divided by the number of days, make a *mean for the month:* and twelve of these latter, for successive months, added up and divided by that number, the *mean for the year.* Lastly, the comparative heat of the place, or *Climate,* is found, by taking in this way the average of a sufficient number of the *means* of successive years. (*o*)

Climates, thus examined, are found to differ by pretty steady geometrical proportions according to the *Latitude,* other circumstances being alike. Here are, marked on the globe I have before me, the calculated Mean Temperatures of sixteen several Latitudes—freezing cold (of course) *near the pole,* and growing warmer as we proceed South; till, *at the Equator,* we find a Mean heat for the year, equal to that of our own summer weather before thunder; —the Temperature rising at one time as much above that, as it falls at another below it. If we begin here then, under the Line, we have a Climate the mean of which is 84° : at ten degrees of North Latitude, we have the mean 82° —at fifteen degrees, 80° —at twenty degrees, 77° —at twenty five degrees, 74° —at thirty degrees, 70° —at thirty five degrees, 66° —at forty degrees, 62° —at forty five degrees, 57° —at fifty degrees, 53° —at fifty five degrees, 48° —at sixty degrees, 44° —at sixty five degrees, 40° —at seventy degrees, 37° —at seventy five degrees, 34°: and for the vicinity of the Pole, or eighty degrees—and within that, 32° : these Temperatures are however to be considered only as *calculated approximations.* (*p*)

While this gradation is fresh in mind, let us proceed to demonstrate its Astronomical causes, and the manner in which the effects are produced. We will place, for this purpose, the globe representing our Earth in that position, with respect to the Sun, which it assumes at the *Equinoxes.* The rays are considered always as falling upon the Earth *in parallel lines*—so that, where the Sun is vertical, (as to the Equator, at this season,) equal measures of surface receive equal quantities of heat. But the Sun is *vertical* only to that particular latitude: or to its parts, as presented in succession through the twenty four hours:—to the remainder of the Earth the rays are more and more oblique, as we recede further towards either Pole: so that equal measures of the surface here receive very unequal quantities of heat. (*q*) Let A B in this diagram represent ninety degrees of *Latitude,* contained in the space between the Equator and the Pole; and the parallel lines, *the Sun's rays.* Then, if we divide the Sun's influence into ninety equal breadths, we shall find that the space between the Equator and Latitude thirty degrees will get for its share forty five breadths, or one half of the whole influence: the space from Latitude thirty to sixty, thirty three breadths, or little more than a third of the influence; —and the remaining thirty degrees, extending to the Pole, only twelve breadths, or little more than an eighth of the whole influence of the Sun, for a full third of the whole extent in Latitude. Thus the parts opposed to the Sun get an excess of heat, at the expense of the parts turned from it.

Even at this season, however, the Sun's influence is carried over and beyond each Pole, by an effect of the Atmosphere on the rays, which is called *re-fraction.* (*r*) Where a ray of light falls vertically on the Atmosphere, it passes in a direct line to the ground: where obliquely (and the more so, as it is more oblique) it is *bent down* in its course, by the attraction of the Earth and Air, and comes to the surface in a curve: hence we have our morning and evening twilight—of which more in another Lecture. For the present it may be observed, that this property in the Atmosphere has the effect of *gathering the Solar influence about the Poles,* to a degree which considerably lessens the extreme cold and darkness of those regions. (*s*)

We have thus far viewed the Climates of our globe as they are found (in respect of heat) *at the Equinoxes.* Let us suppose it to have been the time of the Vernal Equinox, or beginning of Spring, with the Sun in *Aries,* and let us now see *how we get our Summer.* In the course of the Earth's revolution about the Sun, it is brought (at the season we call Mid-summer, but which *begins* it rather) into this position; the Sun in Cancer, (*t*) with the North Pole turned *towards,* the South *from* it. The Sun is now so far brought over our Pole, as to shine upon *it* through the whole twenty four hours; and upon us in these islands for two-thirds, or more, of the time. The consequence is, that we receive a proportionately greater share of the direct influence of the rays; (*u*)—while the Southern temperate regions (our antipodes) turned away from the Sun in proportion, are deprived of it. Thus we get our Estival Season: —but from the time that it is fairly begun, the Earth in going its Annual round is bringing us back again to the *Balance,* (as that sign is called which denotes the Autumnal Equinox); and on arriving at it we are again in the position which agrees with a Mean temperature, (or mean share of the Solar heat), with an equal length of day and night.

What was the Vernal Equinox to *us*, was the
Autumnal to our Antipodes in Australia : now, that
they are getting their Summer, by the change of the
Earth into an opposite position with regard to the
great Luminary, we are proceeding to our Winter.
To represent this, we have only to conceive of the
Sun as removed to the other side (in the sign Capri-
cornus), and shining *over the other pole*, thus. (*v*)
In this position of the Earth, we are but about a
third of the twenty four hours in daylight; and the
North Polar-regions, wholly deprived of day, are
indebted for the degrees of illumination they still
enjoy, to the Moon and Stars, and to the coruscations of the Aurora Borealis.

It would scarcely be needful to repeat the mention
of the Equinoxes, with which we began—were
it not for the state of the regions *immediately under
the Line;* which have now their greatest heat, the
Sun being at noon directly over head.   But even
these are relieved, and rendered habitable, by the
passage of the luminary to either Tropic; to become vertical to the Latitude of twenty three and a
half degrees, North or South, and return.   This
change of position is attended in those regions with
a decided periodical *change of the seasons*, from
dry to wet and the contrary ;—and with corresponding changes of Temperature : of which more hereafter.

But there is a further source of variation of Climate, which operates as universally as the Latitude,
—*elevation above the level of the sea*.   I have said
that the Atmosphere itself is colder the higher we
ascend into it: this is to be understood of its ordinary
state, there being occasionally found in it *currents*,
at a higher temperature than the air below. We may
observe snow on the hills, before and after that

which falls on the plain country; and we know that
on the highest mountains it lies unmelted *through
the year:* and this even under the vertical sun of
the Equator. The heat we experience in Summer
is a compound result, of the direct action of the rays
on the Earth's surface and of the reflection of these from
the ground. On the summits of lofty mountains,
and in general in high tracts of land, the latter condition is in great measure withdrawn: these portions
of the surface, placed in the midst of a cold atmosphere and subject to every current, fail to accumulate
heat by an extended reflection. After considering
these facts, you will be prepared to hear assigned
to elevated countries a Climate, cooler in proportion
as they are higher. Thus *Quito* in Peru, elevated
nine thousand feet, enjoys a mean Temperature of
sixty two degrees, *under the Line;* while at Spanish
Town, Jamaica, (in nearly the same Longitude) but
eighteen degrees North of the Line, they have a
mean heat of eighty one degrees. (*w*)

The Ocean itself, though subject to the variations
on its surface consequent on Latitude, and making
a hot or a cold climate accordingly, *at Sea* as on
Land, is yet more uniform in heat and less subject
to either extreme, than the latter. The reason is,
probably, that the Sun's rays entering further into
the substance of the Sea warm it to a greater depth,
while the constant internal motion of its waters dis-
tributes the heat in the mass: hence the Sea will
cool a very hot atmosphere, and warm a very cold
one. Islands, and narrow prolongations of the land
into the sea, (such as the county of Cornwall)
are hence found to have a more uniform Climate,
as to Temperature, than the larger breadths of land.
They escape frost and snow, and pay for the privilege
in the greater frequency and abundance of rain.

It is on the extended plains of the Northern Continents, and in the vast sandy deserts of Africa, that
we must expect to find exemplified the extremes
of Temperature. Here are the magazines of cold
from which we import our protracted Easterly
Winds: and the birth-place of the suffocating Simoom,
or the sultry *Sirocco* (so named in Greece and
Italy from its coming to them over *Syria*,) by which
the very sand is borne, along with an unhealthful
influence, to vast distances over sea and land. (*x*)
In both cases, an extensive contact of the atmosphere with a plane surface of land brings on the
effects: the *heat*, from radiation downwards—the
cold, from an opposite cause. For it is a curious
fact, now fully established by observation, that after
a hot day in a clear atmosphere, *the soil radiates by
night into space*, losing rapidly in this singular
manner the heat it had previously acquired: so that
both in severe frost and in the deposition of the dew
by night, a Thermometer placed on the ground descends several (at times *many*) degrees lower than
one suspended but a few feet above it. (*y*)

Climate is also modified in other respects, *by the
nature of the surface*. Extensive forests, by excluding the Sun from immediate action on the soil,
tend to make a country cold and rainy: an effect
the reverse of that we find produced by sandy deserts.
Cultivation of the soil tends to make a more *uniform*
Climate; to take down somewhat the heat of Summer, and temper the winter's cold. It is probable
that our own Climate has *a mean heat* pretty nearly
the same with that it was subject to, when the Island
was occupied by the Romans: but there are circumstances in its History, which may make us
suspect that, for some space in subsequent but early

times, when the greater part was yet covered with forest, there existed *sheltered spots of peculiar warmth and fruitfulness*, (now exhausted and laid open) in which even the vine was cultivated on the great scale with success.

More abundant rains about the Equinoxes with more of wind and thunder, drier and somewhat hotter Summers, and more continuous frosts in winter, *may* have constituted the difference of our ancient Climate from the present: so that, without its *basis* being disturbed, changes may have been wrought, tending to a more equable distribution of rain and warmth; and these, effected in a manner so gradual as to have escaped observation. It is reported, by observing persons, that the like changes are now going on in the most antiently cleared and cultivated parts of the *North American Continent*.

It will appear, on examining the Astronomical causes of the diversity of Seasons, that they are calculated to produce the greatest possible effect, with the least possible expense of *power*: they are an arrangement of nature, which no human sagacity could have devised. Had the Earth been presented to the Sun throughout the year, in like manner as at the Equinoxes, the Tropical heat would have been always and alike intense;—and, the rays being equally distributed to the two hemispheres, North and South, the air would at all times have been in a freezing state at the *Poles*.

As it is, though these *extreme points* are from the nature of the surface inaccessible to man, we see them placed alternately in sunshine during half the year. The result is, that the parts around them have a Temperature sufficient for a scanty vegetation, (z) and for the support of an abundance of animal life, both on the earth and in the sea. Hence *mankind* also have obtained a residence, and gain a subsistence, in places where one would least expect to find them: where the watery element is never wholly reduced to a liquid state, but ice encumbers the sea in the midst of summer; lying also under the soil like stone and forming part of the solid country.

Here, seated in winter in hemispherical cabins of snow, lighted by windows of ice—and spending the short summer in tents, wherever the game or the fish are to be found, the little Esquimaux live, content as those who are sheltered by the proud domes of palaces; and happy *in their way,* as a rural population!

Again, the countries between the Tropics become habitable in the present state of things to a peculiar race of men who, could they have endured the constant heat, would yet have been deprived but for this arrangement, of the luxuriant vegetation, and subsequent perfect ripening of the fruits, to which they are indebted for *their* shelter and *their* food. Further, had the abundant Tropical rains not been limited to a certain season, had they been of continual occurrence, it is doubtful whether the climate would have been fit for any thing but giant palms to live in:—whether the constitution even of the Negro could have borne so long the unhealthy influence of a hot and moist air, tainted by the putrid spoils of the vegetable kingdom. But, as things are, man finds here also the necessaries of life, with all the physical requisites for its enjoyment. Such are the benefits which the Providence of God has conferred on us by means of that simple contrivance, the oblique position of the Earth in its orbit: we shall have occasion to trace them further, as regards our own Climate, in the next Lecture.

LECTURE THIRD.

*Cycles of Temperature : increase and decrease of
the heat through the Seasons: mean and
extreme Temperatures of the Years,
Months, and Climate.*

In treating the differing Temperatures of different
Years, and the gradation of Heat and Cold through
the Seasons, it will be convenient to take for our
example the Climate of London, or of the country
in which the Metropolis is situate; because in that
part of the island the *Annual mean Temperature*,
with the Mean Temperature of each *Month* in the
year, and other data needful to a systematic account
of the Climate, have been accurately found. (*a*)

That Year differs from Year as warm or cold, is
matter of common observation—especially with such
as are interested in the results of favourable or in-
clement seasons : but it is not very long since the
proportions in which they differ have been ascertained. From above forty years' observations made
on the Thermometer, in the open air in Westminster,
the station being Somerset-House, and at several
other places about London, it is found that the Mean
Temperature of *the year* varies four degrees and

E

eight tenths in different years: the standard or *Mean of the Climate* being nearly forty eight degrees and eight tenths—in figures, 4·8° —48·8°.

The City is warmer than the country, on the whole year, by 1·579° : and this excess of heat is greatest in the winter and least in the spring. The gradation of the difference between the two, through the several *months*, will appear in the diagram, which I shall exhibit by and bye, of their respective Mean Temperatures: but we must first shew the differences in the Mean Temperatures of a series of *years*. (*b*) Seven years then in London, from 1800 to 1806, have their several mean heats as follows, viz. 1800— 50·522° ; 1801—51·080° ; 1802—50·200° ; 1803— 50·329° ; 1804—51·731° ; 1805—49·998° ; 1806— 52·734°. Here is a gradation *from cold to warmth:* —for the Mean of 1799 was only 47·920° ; which is 4·814° lower than the temperature of 1806. The diagram which I now exhibit shows, by simple inspection, the fact of a gradual addition of warmth in this series, proceeding mostly by alternate years. But the seven years from 1810 to 1816 (after three years about the Mean) show as clearly the opposite gradation, *from warmth to coldness;* and that effected, in like manner, by alternate years—every other year regaining some heat. The temperatures of these years, in the Country, were as follows, 1810—49·507° ; 1811—51·190° ; 1812—47·743° ; 1813—49·762° ; 1814—46·967° ; 1815—49·630° ; 1816—46·572° . The difference between the coldest and warmest year, is here 4·618° : shewing a loss of heat by which this series pretty exactly balances the preceding one. (*c*)

An ascending series of temperatures, mostly rising as before by alternate years, began with 1817 : and

a descending series (after three years of mean heat
as before) with 1827 : so that it is probable we have
in this island a *Cycle of temperature*, in which the
Climate becomes gradually warmer and colder by
turns: in such a way as to exhibit both Extremes
in the space of seventeen years. But, until observations shall have been more extensively examined
and compared, we cannot propose a full and definite
theory on this subject. (*d*)

It is otherwise with the *Seasons*. Here we have
a beautiful System of Temperatures, founded on
observation; which I shall now proceed to expose.
Let the broad circular coloured band, in this diagram,
represent the year of Natural Temperature; divided
first into its four *Seasons;* Spring being tinted green,
Summer purple, Autumn straw colour, and Winter
grey : secondly, into the *Months*, with their names
annexed: thirdly, into portions of these, of five days
each ; forming a *Map* on which may be marked the
boundaries of *the heat and cold of the year.*

Then, let the circle in the middle represent the
Equator, and the dotted line close to it the place in
the scale for the *Mean Temperature* of the Year :
the remaining concentric circles forming a scale of
temperature, from seventy degrees on the inner
circle to thirty degrees on the outer.

Lastly, let the *regular* excentric circle represent
the *Ecliptic*, in its proper relation to the Equator ;
and the *irregular* excentric circle near it, the course
of the daily variation of the Mean Temperature
through the year. Bearing in mind this principle,
that all has relation to the daily average warmth,
or the Mean Temperature (on twenty years) of each
day in succession, let us now proceed to trace the
advance and decline of the heat through the Seasons.

I said in a former Lecture, that the heat we enjoy
in our atmosphere is a compound result, of the direct action of the sun's rays and of the reflection of
these from the ground. The most familiar illustration that I can give you of this, will be found in the
effect of the tin screen, which the cook places before
the fire while the meat is roasting. Get for a moment
between this and the fire,—you will perceive how
much hotter it is there, than at an equal distance
from the fire without the screen ; and will know how
to appreciate the action of a vertical sun upon a
sandy desert!

We are apt to think the close fitted turban of the
Asiatic a singular head dress ; and such as must be
very encumbering in a hot climate ; but this light
yet bulky covering is, precisely, the defence which
experience has taught the wearer to place upon his
head, in situations in which the direct impulse of the
sun might otherwise prove fatal.   And for the rest of
his person, which is exposed to a bath of hot air on
all sides alike, he merely envelopes it in the lightest
and loosest garments he can put on. (*e*) For ourselves,
in this temperate climate, though rarely obliged to
take means for escaping a *coup de soleil*, we are yet
always placed, in the height of Summer, between
a fire in the sun over our heads, and a reflector in
the earth beneath our feet. *(f)*

Beginning our examination with the shortest day,
at the time when the sun is lowest, we might
naturally expect to find the Temperature likewise at
its lowest point:—but it is not so. If we follow the
flexuous line which describes its movements, we
perceive that it descends for some time afterwards
in the scale.   Again, if we inspect it at the Vernal
Equinox, where the sun crosses the Equator, we

find the Temperature not yet in the *Mean ;* which it does not attain till a Month has elapsed. And the greatest *heat*, in like manner, is found to commence a full month after the Summer solstice : and the Mean Temperature of the year to occur, the second time, a month after the Autumnal Equinox. Thus, on the evidence of twenty years' observations, the average daily Temperature is found to follow the Sun at an interval of thirty days, or a sign of the Zodiac, through the year : the coldest season coming on at his entrance into *Aquarius*, the hottest at his entrance into *Leo :* and the Mean Temperature, in Spring and Autumn, coinciding with his entrance into *Taurus* and *Virgo*, The reason of this difference is that the *Earth*, having imbibed heat in the Summer, requires to lose it before the Winter's cold can be established : and in like manner having been cooled in Winter, it requires to be warmed again before Summer can begin.

Let, now, the red Crescent in this Figure represent the quantity, or extent on the scale, of *the total* HEAT *of the year ;* and the blue Crescent, the *total* COLD *of the year :*—the one being the space included between the *higher* temperatures and the Mean ; the other, those betwixt the Mean and the *lower* temperatures. Then, if we wish to see described also the extent on the scale of this borrowing and lending of heat, between the Earth and the Atmosphere, the thin blue Crescent opposite *Spring*, in this other diagram, will shew the *ground-cold* which in that season keeps back the Summer : and the red one opposite the Autumnal Months the *ground-heat ;* which being given out to the air retards the approach of Winter. (*g*)

E 3

The Extremes of Temperature of the CLIMATE, or greatest heat and greatest cold in a Cycle, appear at nearly equal distances above and below the *Mean*. The greatest *cold* of the Year occurs commonly in January: the greatest *heat* is not limited to July, but is found as often in some other month. Of the twelve months, there are only two in Spring, and two in Autumn, in which extreme heat or extreme cold may not at times be found.

The following is a Table of the years from 1807 to 1816, as observed in the Country.

Year.	Highest Temp.	Lowest Temp.	Range.	Medium.
1807	87°	13°	74°	50°
1808	96	12	84	54
1809	82	18	64	50
1810	85	10	75	47·5
1811	88	14	74	51
1812	78	18	60	48
1813	85	19	66	52
1814	91	8	83	49·5
1815	80	17	63	48·5
1816	81	5 minus,	86	38

Extremes 96   Minus 5: Mean 72·9 Mean 48·85

From ninety six degrees to five below zero is one hundred and one degrees; which appears to be the extreme range of the Thermometer in the Climate of London. The *average range* is nearly seventy three degrees.

The greatest heat of these ten years occurred from the twelfth to the fourteenth of July, 1808: the observations were made by a Sixes (*h*) Thermometer, fixed to a short post on a grass-plot, under the shade of a Laurel Tree: five miles East of London. On the twelfth, the heat was ninety two degrees, on the thirteenth, ninety six degrees, on the fourteenth ninety four degrees: so that these three days are the hottest on record in this part of the Island. The heat was not of that sultry oppressive kind common before thunder: the sky was serene, and a fine breeze prevailed. Yet, such was the ardour of the sky, that all motion was unpleasant, and labour in the sun dangerous: the birds were mute all day, and revived by the freshness of the night were heard singing by moonlight. In the evenings, *dew* fell pretty freely; and at a Temperature which in ordinary circumstances would not have admitted of it: but the production of dew depends, not on the absolute but on the relative coldness of the atmosphere. The natural *evaporation* of water being examined, on the 12th, it was found that four hundredths of an inch depth was thrown into the air in an hour, from a vessel placed in the shade: and near an inch evaporated in the three days. So that here was an abundance of moisture in the atmosphere, to be deposited in dew, upon the loss of about thirty degrees of the heat.

This heat, which was very extensively felt in Europe, gave occasion to violent thunder-storms:—not indeed at the place of observation, but near enough for the lightning to be seen on the horizon. One of these of peculiar violence is thus described, as occurring on the night of the fifteenth, in Gloucestershire and the neighbouring counties.

" Unlike the tempest of the milder zones, the thunder was remarked to roll in one continued roar, for upwards of an hour and a half; during which time and long afterwards, the flashes of Lightning followed each other in rapid and uninterrupted succession. But the most tremendous circumstance of this storm was the destructive *hail shower* which accompanied its progress. It may be doubted, however, whether such a name was applicable: for the masses of ice which fell bore no resemblance to hail-stones, in magnitude or formation; most of them being of a very irregular shape; broad, flat, and ragged; and many measuring from three to nine inches in circumference. They appeared *like fragments of a vast plate of ice*, broken into small masses in its descent towards the earth." The damage done by the storm was proportionate to its intensity: buildings were struck, persons and cattle killed by the lightning; the windows and glass coverings of hot-houses were demolished; the trees stripped of their leaves and fruit, and the crops reaped and thrashed, by the hail. (*i*)

A comparison of the variations of wind and temperature, as occurring at the same time at London and Paris, (the observations at the latter being reduced for the purpose to Fahrenheit's scale,) will prove the extensive action of the causes concerned in producing this extreme. The lower curve in this diagram shews the variations of temperature at Plaistow near London, from the tenth to the twenty second of the month: the upper, those at Paris, distant one hundred and eighty miles to SSE. (*k*)

The maximum heat at Paris on the tenth was 82·6°—the wind being NW: at *Plaistow* it was seventy six degrees—wind SW: during the three following days, the heat at each place increased steadily, the wind at Paris E. and SE: at *Plaistow*, S. and SW. On the thirteenth, when the Thermometer with us had risen to ninety six degrees, the evening sky presented a dewy haze to S E. and some traces of thunder-clouds to NW. The change to a lower temperature was therefore beginning to approach *us*, while the atmosphere at Paris remained undisturbed; for the heat rose still with them to the fifteenth, when it was 97·2°, the wind SE. In the mean time, our heat was reduced by a NE. wind to 81° —and this change of temperature was two days in getting to Paris. Their heat was reduced on the seventeenth to 81·5° by day and 62·7° by night. A second elevation of the temperature was next felt for two days, at both places; after which the heat at both went down to the ordinary summer standard; by a SW wind, introducing rain. It appears *on the whole* that Paris, consistently with its more Southern latitude, had about four and a quarter degrees of heat more than London.

The Extreme of *cold*, here presented, in comparative curves for the stations of London and Paris, was observed at Tottenham near London, on the ninth and tenth of February, 1816. *Paris* on this occasion shews an average Temperature exceeding by 4·67° that of London: but the cold extreme is not so low by eighteen degrees; and it occurs (as the hot one did) *two days later*. In the subsequent *rise* of the Thermometer, we see Paris take the lead, contrary to the order in the case of heat; and the cold goes off at both places by a Westerly wind, on the fifteenth of the month. All these circumstances admit of an explanation founded on the respective geographical positions of the two places; as to latitude and relation to the Atlantic ocean. A gentle breeze from an extensive surface of *land* seems requisite, to bring on extreme heat or cold: a more brisk wind from the *sea* commonly removes either.

The following is a Table of the Temperatures observed—those of Paris reduced as before. (*l*)

PARIS.

1816.	MAX.	MIN.	
Feb. 5	45·68°	36·50°	W.
6	48·65	43·25	SW.
7	48·65	32 45	SW.
8	30·20	21·65	NE.
9	22·10	18·25	NE.
10	26·15	15·35	ENE.
11	27·05	12·65	E.
12	35·60	22·55	NE.
13	36·50	22·25	NE.
14	41·00	26·15	N.
15	41·45	33·65	WNW.
16	44·60	39·65	W.

LONDON.

Feb. 5	39	35	S.	Misty
6	38	31	SE.	Rain : sleet
7	31	15	NE.	Deep snow
8	24	7	N.	Clear
9	20	—5	E.	Clear
10	30	19	SW.	Clear: *Cirrus*
11	37	18	N.	Sleet: snow
12	32	11	N.	Clear
13	36	22	Var.	Clear
14	39	25	W.	*Cirrostratus*
15	44	32	SW.	Misty: cloudy
16	47	33	NE.	*Cirrostratus*

Mean of twelve days at Paris 32·17° —at Tottenham 27·50°.

We had at Tottenham, on this occasion also, a clear atmosphere by day : a gale from the NE. had precipitated in *snow* the moisture which abounded in the air, and which had twice brought the Hygrometer to one hundred degrees, or the point of

saturation. So cold was the surface on the ninth
at noon, that a bright sunshine effected no change
in the snow :—the polished plates of which continued
to reflect brilliantly, and shew the colours of the
prism. In the night, the Thermometer which had
risen only to twenty degrees, went down to minus
five : and there is every reason to think that it continued below zero for about twelve hours. It would
have been dangerous to the naked hands, at this
time, to have handled wood or metal out of doors :
a wet finger, placed on the iron railing in front of
the house, adhered instantly by freezing, and the
projecting parts of large keys, left in the locks of the
outer doors, collected rime from the atmosphere
within the house. (*m*)

Such nights are happily of rare occurrence in our
Climate : probably not above five of them fall within
a Century. Not that the first impression of the cold
in a calm air is disagreeable : the extreme dryness
makes it very much a non-conductor, while the
powerful electricity stimulates the skin and lungs :
one may keep warm with exercise at such times in
ordinary clothing. It is during a change of temperature in either direction, and in strong winds with
moisture attending, especially when prevented from
using exercise, that the natives of temperate Latitudes suffer the most. In Climates where intense
cold is certain, the population clothe accordingly :
the person is well defended by a fur covering, and the
winter dwellings are constructed with a due regard
to the maintenance of a sufficient heat within. *Here*
we are too commonly taken at unawares : the poor
shiver alike in the street and in the cottage, the lost
wanderer perishes in the snow—and those most
provided with the means of averting danger suffer,

by imprudent exposure and consequent inflamma-
tory disease. We may possibly, yet, see the time,
when a more perfect knowledge of the subject shall
have enabled us to expect, and put people on their
guard against, the dangers arising from the rare occurrence of such extremes. It may be remarked in
conclusion, that our greatest heat and severest cold
appears to give place, after prevailing through a
comparatively calm season, to the irruption of the
more temperate air of the Atlantic : and that the heat
given out by the vapour contained in our atmosphere, condensed in rain and snow at the approach
of the Northerly current, must tend considerably to
abate the rigour of the frost. (*n*)

We have gone through the account of the Mean
and Extremes of the *Climate*, and of the *Year:*
those of the several *Months* we shall review presently.
For *day* and *night*—the difference of temperature
between them varies much in different seasons:
sometimes not five degrees, (as in continued cloudy
weather and during moderate frosts ;) at others, the
difference amounts to thirty degrees or more : the
latter is most frequently the case in clear weather in
spring:—and it is then that we must use especial
care, to defend the bloom of our trained fruit trees
from night frosts. This is more easily done than
would be supposed : the loss of heat into a clear
air, by night, is chiefly from the *radiation* to the
sky above—hence if we *intercept* (so to speak) *the
view of the sky* from the fruit wall, that degree of
protection suffices, although we put nothing on it
that shall seem to keep the trees *warm*. A very
open *net*, more especially of *woollen* yarn, will be
found preferable for this purpose to close matting :
and some have thought it sufficient to interpose be-

tween the trees and the sky a kind of narrow pent-
house of straw, on the top of the wall. (*o*) The *mean
difference between day and night*, for the year, is
about eleven degrees in London, and fifteen in the
country: the obscurity of the atmosphere of London,
in calm weather, is unfavourable to a free radiation:
hence it is in the *nights*, chiefly, that we find the
superior warmth of the city, making a less difference
of course from the heat of the day. The variation
of the Temperature from day to night is greatest in
the longest days, and least in the shortest: we shall
treat of its *Monthly* averages presently: but we have
now to perform our last operation on the Map of the
Seasons, and shew in figures their limits and proportionate temperatures. (*p*)

If we cut the *red* Crescent, or warm side of the
Map, at fifteen days before the Summer solstice, and
again at fifteen days before the Autumnal Equinox,
we shall have the limits of the *Natural Summer*; here
distinguished by a purple tint, and comprehending
ninety three days from the seventh of June. In this
space lies the bulk of the *heat* of the year. And if we
divide the *blue* or cold side of the Map, at fifteen
days before the Winter solstice, and reckon from
thence eighty nine days (or in Leap-year ninety, beginning the seventh of December) we shall have the
Natural Winter, comprehending the principal *cold*
of the year; coloured grey.

We have now left the intermediate spaces, determined by the limits of the other two: shewing
Spring, coloured green, comprehending ninety three
days from the sixth of March: in which space the
cold wears out, and the lower Summer temperature
is established. And *Autumn*, including ninety days

F

from the eighth of September; during which the Temperature (crossing the Mean line in October) is reduced to the cold which begins Winter.

The rise and fall of the daily Mean temperature is thus shewn to be symmetrical through the year: Spring beginning with 39·96° and rising to 58·08°, we gain 18·12° of heat in this season: Autumn, falling from 58·16° to 39·96°, there is a loss of 18·20°, balancing the rise in Spring. Again, Summer ascending 6·87° to 64·95°, and descending 6·79° to 58·16° — and Winter descending 5·51°, from 39·96° to 34·45°, and rising 5·51°, to 39·96° with which we began, there is a perfect balance also in these two seasons.

But the symmetry is not less striking, if we compare the *Mean temperatures* of the four; or their average heat, found by adding together the Mean temperatures of the days, and dividing them by the number of days: for we find 48·94°, the Mean of Spring, to exceed by 11·18° that of Winter, which is 37·76° — and 60·66°, the Mean of Summer, to exceed by 11·72° that of Spring: and 49·37°, the Mean of Autumn, to fall short by 11·29° of that of Summer, and 37·76°, the Mean of Winter, to be 11·61° lower than that of Autumn. Thus the four seasons, divided according to their natural limits, ascend and descend in the scale of warmth by equal proportions:—a System resulting indeed from observation alone, but approaching perhaps as nearly to the boasted precision of Mathematical science, as in the present infant state of Meteorology (for it is but as the Astronomy of the Chaldean shepherds) we have a right to expect. (*q*)

Let us now finally review the Months in their order, as placed in this diagram: with their proportionate Mean and Extreme temperatures, and other particulars belonging to this part of the subject. (r) The *dotted curve* in the Figure shews the Sun's progress in declination.

January has, in the middle of the month, about eight hours and twenty minutes of the Sun's presence above the horizon. The Thermometer rises, on an average of twenty years (ten of them in the City) to 40·28° in the day, and sinks in the night to 31·36°: difference 8·92°. Consistently with the average lower temperature, the greater part of the nights in this month are frosty. Mean of the Month 36·34°.

February has, on a mean, nine hours fifty five minutes of daylight. The mean of temperature by day is 44·63°; by night, 33·70°: difference 10·93°. *Eleven* nights in this month, on an average, are frosty: Mean 39·60°.

March has in the middle of the month eleven hours fifty minutes of day. The average Temperature rises to 48·08°, and falls to 35·31°: difference 12·77°. The average of frosty nights in this Month is *twelve*. The month was coldest (as to the period of twenty years examined) in 1799 and 1807, and warmest in 1801 and 1815. Evaporation is now commonly very brisk, from the great prevalence of Northerly winds, and tends to keep down the Temperature; as is apparent from the depression of the curve in this part. Mean 42·01°.

April. Mean daylight about thirteen hours fifty seven minutes. Average temperature by day 55·37°, by night 39·42°: difference 15·95°. Warmest in 1798 and 1811, coldest in 1799 and 1808. Frosty nights on an average, *six:* never missed, in ten years from 1807 to 1816, to have some frost in the month, night or morning. Mean 47·61°.

May has, in the middle of the month, fifteen hours thirty five minutes of daylight. The heat rises on an average to 64·06° and sinks to 46·54°: variation 17·52°. It was warmest in 1804 and 1811, and coldest in 1802 and 1816. In five seasons out of ten, this month is subject to a frosty night or two: and these are the more mischievous to the crops from being so few. Mean 55·40°

June. Length of day in the middle of the month sixteen hours thirty two minutes. Thermometer rises by day, on an average, to 68·36° —falls by night to 49·75° —difference 18·61°. This month had no frost, in the ten years from 1807 to 1816. It was warmest in 1798 and 1811, and coldest in 1797 and 1812. Mean Temperature 59·36°.

July. Length of middle day sixteen hours five minutes. Mean of higher temperature 71·50° —of lower 53·84°: difference or variation 17·66. The month was warmest in 1803 and 1808, and coldest in 1802 and 1812. Mean temperature 62·97°. No frosty nights, in this Latitude, in the month.

August. Has the sun in the middle of the month for fourteen hours thirty two minutes. The Temperature rises on an average to 71·23° and sinks to 53·94°: variation 17·29°. It was hottest in 1802 and 1807, and coolest in 1799 and 1812. This month has no frost, in the Latitude of London; and shews the warmest *nights* of any in the year. Mean temperature 62·90°.

September. The middle has twelve hours thirty nine minutes of daylight. The average greatest heat by day is 65·66° —of cold by night 48·67°: variation 16·99°. It was warmest in 1804 and 1810—coldest in 1803 and 1807. We have now, occasionally, a frosty night. Mean Temperature 57·70°.

October. Middle day ten hours thirty seven minutes long. Greatest average heat 57·06° —average cold 43·51°: variation 13·55°. The month was warmest in 1804 and 1811—coldest in 1797 and 1814. It is less subject to frosts by night than we might expect from its place in the curve: but the condensation of the summer vapour, which now takes place, giving out its heat to the medium in which the water is deposited, serves to keep up the temperature of the nights and prevent frost. Only four nights on an average are subject to it. Mean temperature 50·79°.

November. The middle of the month enjoys only eight hours forty nine minutes of day. The average temperature rises to 47·22° and sinks to 36·49°: difference 10·73°: coldest in 1798 and 1816—warmest in 1806 and 1811. About *twelve* nights on an average are frosty; the curve presents a sensible depression in this part, answering to that in March: Mean Temperature 42·40°. In the years from 1811 to 1816, in a series proceeding to the lower Annual temperature, the frosty nights in November shew the following gradation of numbers: 5, 9, 12, 14, 17, 20.

December. Day in the middle of the month, reduced to seven hours forty six minutes. The average temperature by day rises to 42·66, by night sinks to 33·90°: difference 8·76°. It is the month in which the variations of the Temperature depend least on the Sun's presence (which is so short a time above the horizon); and most on the *wind*, as Northerly or Southerly. On an average of ten years, about half the nights are frosty.

We see, by this account of the difference between day and night, that it proceeds like the Mean Temperature by a pretty regular gradation, following the

altitude of the Sun: the diagram now exhibited
shews its progress through the year—the dotted line
being that of the declination *reduced to half-scale.* (s)

These Results are founded on observations, one
half of which were made in the City. The warmer
temperature of the Metropolis, (of which mention has
already been made) is a circumstance more or less
applicable to the case of Cities and towns generally;
especially where Manufactures are carried on, requiring the use of fire. It may be enquired, whether
the population itself contributes in any sensible
degree to this effect.

That the superior temperature of the bodies of
men and animals is capable of influencing in some
small degree that of a town, or even of a tract of
country thickly inhabited, will scarcely be denied.
Whoever has passed his hand over the surface of a
glass hive, whether in summer or in winter, will have
perceived, perhaps with surprise, how much the
little bodies of the collected multitude of bees are
capable of heating the place that contains them.
Hence, in warm weather, we see them ventilating
the hive at the entrance, using a peculiar motion of
their wings: and they often prefer, while unemployed,
to " hang out " and lodge, like our citizens " in the
country air !"

But the proportion of warmth so induced must be
small indeed compared with that emanating from the
fires. These are kept, in most of our apartments,
for the sole purpose of preventing the heat which is
generated in our bodies by the process of respiration,
and which raises the blood to ninety six degrees,
from passing *sensibly* into an atmosphere of lower
temperature : a purpose which is scarcely effected if
the air of the room be not at fifty five or sixty degrees.

A temperature equal to that of Spring being thus
kept up within the houses, we have to add the heat
diffused from founderies, breweries, all works that
employ the power of steam, and in general all manu-
facturing and culinary fires. When we contemplate
so many sources of heat in a city or town, the real
matter of surprise should be, that the difference from
the open country is not more considerable. In particular situations indeed, and at particular times,
this excess of heat may amount to ten degrees or
more—but such partial observation is not a basis on
which the calculations of the Meteorologist can be
safely founded.

The use of a Thermometer in the house—and,
(where a convenient situation for it can be found)
also without doors, is not merely an addition to our
store of rational amusement:—it is, with such as
pay regular attention to its indications, a source of
instruction, and a warning to prepare for extremes
of heat and cold. The best construction of one, for
the purposes of a Register, is that invented about
1780, by *James Six* of Canterbury; and still known
by his name. In this Instrument, the moving power
is a column of *alcohol* (too often indeed of ordinary
rectified spirit) included in a thin glass tube, where
it readily expands and contracts in its volume by
the smallest addition or subtraction of heat. This
tube connects with an inverted siphon of smaller
diameter (*t*) containing *quicksilver*: the use of the
latter being merely to serve as a moveable stopper,
and carry the *indexes*. These are short steel wires
capped with glass, and made capable of resting in
any part of the tube, to which they are moved, by
means of a small spring made of horse-hair; or (which
is much better) of a fine glass thread. Into the

upper part of this inverted siphon, opposite the one connected with the spirit tube, is put also a little alcohol; to facilitate the motion of the index on that side.

Thus arranged, the quicksilver being pressed down in one leg of the siphon, by the *expansion* of the alcohol, rises in the other in proportion, and carries *up* the index; which rests at the point to which it has been moved, indicating the maximum of *heat* since it was last adjusted. On the contraction of the spirit, the quicksilver rises again in the other leg; indicating the greatest degree of *cold*. Thus by adjusting the Indexes at nine in the morning, (after noting the place of each) with a magnet applied outside, we secure a like automatic registry of the maximum and minimum for the next twenty four hours.

Such a Thermometer should he screwed to a support, attached to a small post fixed in the ground; an interval being left between the back of the instrument and its support. The *aspect* should be as nearly *North* as possible; and it should be defended from the direct action of the Sun, by some kind of shade on the South side: there may also be a shelter provided on the top against rain:. but this ought not to be so large as to interrupt the radiation to the sky.

The maximum and minimum temperatures of each day, secured in this manner, should be set down immediately in their respective columns of a Register book prepared to receive them, with Notes of the Barometrical variations, the prevalent winds, and the amount of Rain found in the guage. I have described already the manner of deducing the Mean of the Month, Year, and Climate, from the daily observations. (*u*)

There are circumstances attending the congelation
of water and its return to the liquid state, which, as
a beautiful part of the Economy of nature, we must
not omit to notice. The first to which 1 shall advert
is, the temperature at which water acquires its
greatest density, or *becomes heaviest in proportion
to its bulk.*

Liquids (as was before shewn) universally contract
in bulk in cooling: but in the pure element we have
a peculiarity in this respect. It contracts as far
down as the Temperature of 40°—after which it expands, and becomes lighter, by every addition of
cold, down to 32° or the freezing point. (v) When
therefore a pond is about to freeze, the coldest water,
instead of going to the bottom remains on the surface,
and a considerable fund of heat is thus preserved in
the water beneath, which impedes its conversion into
ice. On the other hand, when ice has been formed
to a considerable thickness, and a thaw is coming on,
the water at 40° subsides constantly below the colder
water, and imparts its heat to the ice on which it
stands; thus promoting the speedy return of the
whole to the liquid state.

The second class of effects to be noticed depends
on what is called " Specific heat." It is a law of
nature, that bodies in changing their states change
also their relation to the element of heat. So that
whether we regard this as *matter* or as *motion*, in
its transfer from one substance to another, and in its
continuance within. them, different substances (or
the same body in different states) may be proved to
contain, at equal temperatures, *very unequal quantities* of heat. Thus equal weights of water at 100°
and of quicksilver at 50°, being well shaken together
for a few moments and then set down, the water

coming to the top will be found at 88° —and the quicksilver at bottom at the same. Again equal weights of water at 50° and quicksilver at 100° being thus treated, the temperature of each is found to be 62°: whereas in both experiments, *with homogeneous bodies or those in the same state*, it would have been 75°. The same quantity of actual heat, therefore, that raises the *temperature* of water one degree will raise that of quicksilver 3·16°.

In the *freezing of water* there is a change, not of the substance indeed, but of its *state;* which is attended with like peculiar effects. Water has in it besides the heat of temperature, another and a larger portion of *specific or combined heat:* so that ice at 32° cannot become water at 32° without first absorbing this portion of heat (which would raise the *water* if applied afterwards from 32° to 172°) to constitute it a liquid. And water at 32° must, again, lose the like quantity of heat, by a further and more gradual process, in order to become ice. (*w*)

Thus there is provided a great and universal impediment to the rapid formation of ice by any degree of natural cold.

The lighter water at the surface, cooled down to 32° and having under it water at 40°, has yet to part also with its *specific heat* to the cold air, in order to become solid: and the solid ice cannot become water again, until it has absorbed by insensible degrees the like quantity of specific heat, to constitute it water even at 32°. Each process becomes thus, necessarily, a slow and gradual one. (*x*)

Were it otherwise, the water of ponds, lakes, rivers, nay of the sea itself in certain latitudes, would be liable to become solid throughout its whole bulk in a night's time; upon cooling but a degree below

32°. The effects of this upon navigation, and upon the vast stores of animal life which the waters contain and nourish—the danger, and hindrance and destruction, that would ensue, it would be difficult to estimate.

Again, were it otherwise, the vast *glaciers* contained in the hollows of mountains, and which feed large rivers in the upper country through the summer, would melt (not gradually as now, but) at once, and descend in resistless floods, aided by the snows of the summits and of extensive plains ; to the wasting of every product of vegetation, and the destruction of every work of human art and industry in their way.

It is not needful here to speculate on the means, by which the Creator might have provided for the subsistence of the inhabitants of the waters,—or man, his vicegerent, for the preservation of his own works, on land or sea, in such a state of things. Let it suffice that, *under present arrangements*, we find both the cold of winter and the heat of spring to be manageable influences, subject in their effects to our controul. The fish remain living and vigorous, after having been sealed up, as it were hermetically, in the water of our ponds for many weeks together. (*y*) We are able to break the ice on our canals and prolong navigation, through ordinary winters, in the unfrozen water. Our whalers and our discovery ships (enterprises more fruitful to Science and humanity, both, than the multitude imagine) push into the Northern ocean the moment the ice opens, to pursue a useful occupation—or fearlessly choose their position, for a nine months' residence in ships become convenient dwelling-houses, on a vast plain of ice of seven or eight feet in thickness, and buried in snow ! Nothing of this kind could be done, were not the processes, by which the universal element becomes solid and again liquid, subject to such laws as are here described. (z)

We have, then occasion once more to admire the facilities afforded us, to a right use of the Creator's gifts—thus wonderfully secured by his Providence in Nature's fixed laws: and to acknowledge the goodness and magnify the name of Him who is Lord of all!

—————

ERRATUM:—Page 41, Virgo *read* Scorpio.

LECTURE FOURTH.

The Barometer, its principle, construction and variations; relation of these to the weather and seasons: Evaporation and the Hygrometer: Rain, and its proportions in different seasons, &c.

━━━━━━━━━

In our First Lecture it was stated that the Atmosphere presses on the Earth's surface with a weight of fifteen pounds to the square inch: and it was shewn, that without this pressure we could not raise water by the pump. I shall now demonstrate that the same pressure is capable also of raising quicksilver, though thirteen times heavier than water, to a height proportioned to its specific gravity.

I shall dip this glass tube into quicksilver, and pump out the air from the tube by means of an exhausting syringe. You see that, in proportion as I take off the pressure from the surface *within* the tube, that surface rises. It is a case precisely analogous to that of water in the common pump: but with this difference, that whereas the water follows the " suction," as it is termed, to the height of above thirty feet, the quicksilver can be raised only about as many inches. (*a*)

G

It was shewn, again, that the spring or *elastic power* of the air is equal to the pressure of its own mass: and when this weight was taken off from a portion of air, inclosed in a bottle, the air was found capable of sending up a column of quicksilver to the actual height of the Barometer at the time. We will now repeat this interesting experiment with the help of the syringe—which is in itself a small air-pump. (*b*) The result proves, to demonstration, that the elasticity of the air and its gravity, or pressure, are *Natural forces which counterbalance each other.*

Having gone through these preliminary operations to shew the principle of the Barometer, I shall now proceed to construct one extemporaneously. I shall fill a tube (closed at one end) with mercury, and invert the open end into a cup of that fluid. The tube is purposely made longer than thirty one inches, or the greatest height to which the Barometer rises. You observe that the quicksilver, after descending in the tube, settles at a certain place in its upper part: this point, if I have perfectly excluded the air, will be found at the height of the mercury in any good Barometer at hand. The column has left a *vacuum* exerting no pressure and no spring on the mercury, in the upper part of the tube within; and it is exposed *without* to the full pressure of the Atmosphere: it is in the situation of quicksilver, drawn up by pumping out the air, and then secured in its place by sealing the tube.

You have now before you, therefore, divested only of its appendages the frame and scale, the *Common Mercurial Barometer;* that Instrument so much looked at and so little understood; of the variations of which we are now to treat, as connected with the weather and seasons.

It was invented by Torricelli in 1643: Galileo had before suggested, that quicksilver ought to rise in a pump to a height proportioned to its gravity. This fact Torricelli demonstrated in the simple manner now exhibited; and a third eminent philosopher, *Pascal*, shewed, by carrying the filled tube to the summit of the Puy-de-dôme, a mountain in France, that the mercury was in true counterpoise with the portion of the Atmosphere above it; since it subsided in the tube, as the mountain was ascended by M. Perrier and himself; and rose again as they descended into the plain.

A stationary filled tube, being found to be affected in like manner by changes in the weight of the air, corresponding with certain winds and weather, it became an Instrument for predicting and judging of these, and got at length into common use under the name of " The Weather-glass."

The extent of its variations in this way was finally ascertained; and the scale fixed for ordinary situations at the length of *three inches*, extending from twenty eight to thirty one inches. But for places situate far above the level of the sea, it is necessary to carry the graduation lower, as to twenty seven or even to twenty six inches; the upper part of the scale being shortened in proportion. And Barometers thus graduated, and made portable, are now in common use, to perform the very operation by which the theory of the variation was first proved,—to measure, by the proportionate descent of the column, the height of mountains and other summits, above the sea or any other certain level, below. (*c*)

At the level of the sea, and within the Tropics, the average height at which the mercury stands may be taken at thirty inches English: in the vicinity of

London (at eighty one feet above low-water mark in
the Thames) on the twenty years ending with 1816,
it has been found to be 29·865 inches. The mean
height for the year differs considerably from year to
year: it is higher in proportion as the seasons are
calm, and free from rain and tempests. It is even
found to have regular periods of increase and decrease.
Thus, at Tottenham near London, on nine years
from 1823 to 1831, the averages for the year ran,
29·763—29·878—29·987—30·033—29·938--29·814
—29·688—29·671—29·653; shewing a gradual *increase* in the weight of the Atmosphere of the
Southern part of this Island, proceeding through
four years, by which ·27, or a quantity equal to about
a quarter of an inch, of mercury, is gained; and a
gradual *decrease* through five years, by which ·38
in. or about a third of an inch of mercury, is lost:—
after which the increase began again. (*d*)

The average height again, for any one year, differs
in the different *seasons*. On a careful examination
of the daily observations comprised in one hundred
and twenty four successive Lunar revolutions, from
10th of December 1806 to 11th of December 1816,
it was found that the Winter quarter, beginning
with the Solstice, had gained upon the Autumnal
quarter, beginning with the Equinox, 021—that
the Spring quarter had gained upon the Winter ·030
—that the Summer had gained upon the Spring ·045
in. But in the *Autumn* the whole increase
went off again, the Barometer averaging in this
quarter ·096 (or near a tenth of an inch) below the
Summer. *The column, then, stands highest in the
latter part of the Summer, and lowest at the
beginning of Winter.* We shall have to enquire
into the cause of these differences in a further part
of this Lecture. (*e*)

As the height of the Column, so likewise the range of it differs greatly at different seasons. In the years from 1807 to 1816, the average range was greatest in January and December, and least in July. The following Table, and the diagrams I shall exhibit along with it, shew that the increase and decrease of the range through the year take place with the greatest regularity—if we have regard to the mean effect in the same month of a sufficient number of years.

Average Range of the Barometer near London, in each Month for the years from 1807 to 1816.

	Inches.		Inches.
Jan.	2·38	July	0·99
Feb.	2·01	Aug.	1·02
Mar.	1·80	Sep.	1·54
April	1·62	Oct.	1·82
May	1·52	Nov.	2·12
June	1·25	Dec.	2·37 *(f)*

For the period of seventeen years, extending from 1815 to 1831, the winter range is much greater, reaching to three inches: for the Column stood at 30·80 inches in December 1827, and was down to 27·80 in Dec. 1821: but the summer range is here also greater—the least amount being 1·42 inches for *June;* and not as before, 0·99 in. for July.

The Range, again, is greatest in the North temperate latitudes—we have seen that it attains to three inches in this Island: it diminishes somewhat in Climates much further North, and greatly more as we go *South* towards the Equator: where it is thought not to reach to half an inch at the level of the sea. In high situations the range is also greatly curtailed. For examples of these differences, at Calcutta in N. Lat. twenty two degrees, it is half an inch: on the Missisippi, in Lat. thirty one degrees, it is about an inch; at Rome, in Lat. forty one degrees, an inch and a quarter: but at *Geneva*, Lat. forty six degrees, and *one thousand two hundred feet above the sea*, it falls within an inch. (*g*)

In proceeding to explain the causes of these phenomena, we must have regard in the first place to general principles. It is plain that the Mercurial column is in counterpoise, not merely with the *air* of our atmosphere, but with all that it holds in suspension of a heterogeneous nature: smoke and vapour, mists, and exhalations of all kinds, must be taken into the account—whatsoever the air supports must be concerned in the support of the column. We see, then, why the average height should be greater in a Tropical than in a Circumpolar latitude; and greater in Summer than in Winter. It has here to bear, not the weight alone of the Atmospheric air, but also that of the whole quantity of water which, by virtue of the heat, it is capable of holding. We see also why the Column, after having gained weight in an increasing ratio through the Winter, Spring and Summer, (when it approaches the Equatorial height,) loses all again in the Autumn; the surplus of our indigenous vapour being then disposed of in rain. (*h*)

Again, the column, though on the great scale in counterpoise with the Atmosphere, is not probably at any given hour *an exact measure of its pressure, considered as extending from the surface downward to the Earth.* While changes are being adjusted above, in respect of its weight and motion, others take place below, in respect of its spring and density: and the instrument is found indicating sometimes the one sometimes the other.

To come, now, to the consideration of facts regarding the variations in the height of the column, —*all the great elevations of the Column in this part of the world appear in connexion with Northerly winds*—all the great depressions, with *Southerly.* (*i*) And the Barometer closely watched proves as good an index, at times, of these changes as the Vane itself. The more we enter (as it were) into the Northern Atmosphere, by the continuance of a Polar wind, the higher the glass gets, and the surer the prospect of settled weather. On the other hand, the more we are invaded by the Equatorial air the lower the glass, and the greater the prospect of Rain. The Northern air is dry and dense, it acts as well by its spring as by its weight : in the Southern the case is reversed—it brings us *vapour* (which diminishes the density and spring of the air, in proportion to its abundance :) and this vapour, once introduced into these latitudes, must come down somewhere, condensed by our cold, in showers or in continued rain. (*k*)

A steady current from the North, acquiring heat and taking up water as it comes on, increases the weight of the Atmosphere to *windward :*—a Southerly gale, or a succession of them, losing water by precipitation to windward, makes the Atmosphere lighter on the Barometer:—and both effects, like the swell of the sea before a storm, give indications of the weather that is to ensue.

What effect we may ascribe to a *Centrifugal force* acting on the air, in its violent lateral movements in storms of wind, in lessening the downright pressure, is a subject yet to be enquired into. There is no reason to doubt that we owe to this cause some part of the depressions of the column, which ensue on

such occasions. The meeting, again, of winds
from different quarters, blowing over a great extent
of country, undoubtedly tends to raise the Barometer;
and the drawing off of the Atmosphere of the place
of observation in different directions, to fill the voids
created elswhere by partial rarefaction, to depress
it. (*l*) The very sudden changes that take place at
times, from a rising to a falling Barometer, and *vice
versa*, may be accounted for by the supposition of a
Northerly and a Southerly current crossing each
other, *and at the same time shifting their place
in Longitude.* We are thus taken out of one current and plunged into another, as if by our own
change of place :—while the Vane shifting at the
moment indicates the true cause.

One of the most puzzling indications to an ordinary
observer of this Instrument is, *Rain with a rising
Barometer.* We must consider, here, that Rain is
not often brought to us in vessels ready filled, to be
poured out upon us as they arrive. We find the
Clouds indeed denominated " the bottles of heaven,"
Job xxxviii. 37 ; and there are in the heathen poets
many expressions of like kind : but the fact is that,
in ordinary cases of continued rain, the water is
first brought to us *dissolved in the clear air.* No
finer days are seen, than some in which *with a
falling Barometer* this importation of vapour is taking place: and these, from the frequent change to
wind and rain immediately after them, get in some
places the name of *weather-breeders.* In the Climate of the South of our island, it is commonly a
South-East wind that brings the vapour: and a
North-West that, crossing the other, condenses it.
Of a kind of Clouds, which actually bring rain hail
or snow with them, we shall speak by and bye.

Again, it happens in winter, that a *rise* in the
Barometer, which seemed to promise fine weather,
(and would in summer have brought it in,) is followed
by *rain*. In this case, it is presumable that currents
from the North and South of the place may be
accumulating air over it, by pressing obliquely on
each other; their mixture producing the rain. A
fall on the contrary, in winter, commonly introduces
a reaction from the North, with increased *cold*. (*m*)

The far greater part of the more considerable
movements of the column are found to be, after all,
periodical, and of the nature of tides; being
governed, like the tides of the Ocean, by the Moon's
place, in her orbit or in declination—and most, by
the latter. But of these, and of a daily tide in the
atmosphere governed by the Sun, we may speak with
more advantage at the conclusion of the Lecture.

Evaporation, or the drying up of water by the
action of the air, is one of the simplest of natural
processes; and at the same time one of those the most
conducive to our health and convenience. Few persons reflect sufficiently on the necessary consequence
of it, to fill the air with dissolved water and *make it
moist*: they are apt to conclude that heat alone is
required, to effect the drying of water out of any
substance; and to forget that a medium or solvent
is also needful, to carry it off.

The quantity of water every day thus poured into the atmosphere, from the earth and sea, exceeds
common calculation. It is demonstrable however
from the nature of things, that on the whole average
of seasons and situations it must be equal to the
Rain that falls—*the one being the source of the
other*. A natural process of this kind, alternately
producing vapour and rain, seems to me to be finely

described in the civ Psalm, ver. 6—14 ; and more particularly in the eighth verse : " They [the waters] go up by the mountains, they go down by the valleys, unto the place which thou hast founded for them."

The rate of the production of vapour is according to the heat of the air, and its motion, taken conjointly. A high temperature enables the atmosphere to hold more water—a brisk motion changing the surface, causes it to be more easily taken up.

There are few days in the year, in our Climate, in which some Evaporation does not take place: those occur commonly in winter, at the first approaches of frost : and in dewy nights the process is also suspended; *but it does not often stop during rain.* The calm which attends a change of wind lessens the rate of Evaporation—so also does a moist wind, though it bring no rain. The most intense *cold* will not stop the process : ice evaporates freely into the clear air ; and snows often disappear in a few days, when not considerable, by being dried up without melting. (*n*) A decrease from day to day in the rate of the drying up of the water, is a sure indication of Rain approaching : after rain, the Evaporation commonly goes on for a while at a rate considerably increased.

The common rate per day from a surface of water exposed to a free air is, in winter from a tenth of an inch down to an hundredth : in summer from two to three tenths. The following was found to be the average Evaporation, for each month, in the neighbourhood of London, on the years from 1807 to 1815 : it is undoubtedly greater than would take place from an equal surface of soil or herbage. (*o*)

Inches.		Inches.
Jan. 0·832		July 4·111
Feb. 1·643		Aug. 3·962
Mar. 2·234		Sept. 3·068
April 2·726		Oct. 2·208
May 3·896		Nov. 1·168
June 3·507		Dec. 1·112

Total for the year 30·75 inches.

We may observe, in this Table, how the average amount of Evaporation is modified by the state of the air : being in other respects more or less according to the *Temperature.* For in March, (mostly with brisk Northerly. winds) at a Mean temperature of forty two degrees, there dries away 2·234 inches of water: but in November, at nearly the same Mean Temperature (with fogs and Southerly gales) only 1·168 inches. Again, as to the four seasons, in the three *Winter* months we have, Mean Temperature 37·20°, Evaporation 3·587 inches : in the three of *Spring*, Temp. 48·06°, Evap. 8·856 : in the three of *Summer*, Temp. 63·80°, Evap. 11·580 : in the three of *Autumn*, Temp. 49·13°, Evap. 6·444 inches. If we compare these quantities with the results which should have appeared, had the *heat* alone been concerned as a cause, we shall find that in Spring the effect is *augmented* by about a sixth, and in Summer by more than a fourth part, through the dryness of the prevailing winds : in Autumn *lessened* by more than a sixth, and in Winter by considerably more than a third, in consequence of their dampness. This quality however, when the Temperature keeps up and it is calm, confers on our Autumnal atmosphere a delicious *softness*, comparable to the Climates of the South; which indemnifies us for the keen blasts of the Vernal season.

The best Instrument for ascertaining the Evaporation from day to day, by mere inspection, consists of a shallow metallic cistern, with a scale in it of three diagonals, engraved on glass, in the manner represented in the figure under Note *(p)*. The divisions of the scale should be one tenth of an inch apart—and the descent, in the proportion of one hundredth of an inch to each division. The scale here represented will require a plate of glass, six inches long and one and a half inches wide, to engrave it on: the cistern may be of any dimensions deemed convenient, the proper supports for the scale being only accurately adjusted in it. There is also a small Cistern and cylindrical glass guage, commonly sold with the Rain-guage, for this use. *(q)*

The *Hygrometer* is useful to every Meteorologist; as an index of the actual state of the air at any given time, *in respect of its moisture:* the Farmer also might often avail himself of its indications;—as he would find his cocks of hay, or his sheaves of corn, in the condition as to dryness which this little monitor would shew him before setting out for the field. *(r)*

Daniell's is the most perfect Hygrometer we have; but it requires an operation with an evaporating liquid, every time it is observed. The following are the average indications of De Luc's whalebone hygrometer for every month in the year: the graduation is from 0, in quick-lime, to 100, in a jar over water. Jan. 80—Feb. 75—Mar. 67—April, 60—May, 57—June, 52—July, 52—Aug. 52—Sept. 64—Oct. 71—Nov. 80—Dec. 80°. Here we see, again, the greater dampness of November, at an equal temperature with March.

An interesting branch of this study is, the ascertaining of the *Dew-point,* (sometimes called the vapour-point) or the temperature, as compared with the actual temperature of the free air, at which the vapour of the latter will be condensed into dew. In the evening this is, of course, the temperature of the air itself: in the day-time it is usually several, sometimes many, degrees below it—but if the two are found to approach, *it is an indication of approaching rain.*

I have here a glass, with some water in it newly drawn from the pump: we shall see presently, whether we cannot perform on it the operation in question. You observe, that (as usual when cold water is brought into a warm room) there is an appearance of dew on the outside of the glass. I shall wipe this off repeatedly, examining the temperature as I proceed, with a thermometer having a very small bulb. At the precise point of temperature at which the dew ceases to appear, the air of this room would remain transparent: at any temperature below, with the vapour in it which is now present, it would be filled with visible steam; in other terms a cloud would form in it, to settle in moisture on the walls. When we do this in the open air, and in the middle of the day, the nearer the *dew-point* to the temperature, the nearer we are to rain. The very instructive Meteorological Essays of Dr. Dalton may be consulted, by such as incline to study this part of the subject. The dew on the inside of the windows of our apartments is a deposition of this kind, consequent on the action of the cold without on a vaporous air within. But it sometimes happens, when the temperature of the apartment in a season of frost has become low enough, that a Southerly wind bringing a thaw shall deposit the moisture on the *outside* of the sash, while the

glass remains dry within. An inspection of the
Thermometer, within and without the house, will
in this case at once disclose the reason of the appearances being thus reversed.

Rain, to which we come next in order, is a
fruitful subject; and I shall not be able to tell you
in the remainder of the time the half of what belongs
to it:—but we will give some general results.

The *mean annual depth of Rain* in the Climate
of London is twenty five inches: in some years it
proceeds in excess to thirty two inches: in others it
falls off to eighteen—the former, then, is the rain of
an extreme wet year; the latter, of an extreme dry
one. In a course of thirty five years examined for
this purpose, from 1797 to 1831, the wet years or
those considerably above the mean, are as follows:
1797, (30 inches) 1804, 1805, 1806, 1810, 1812,
1814, 1816, (32·37 inches), 1818 (though an extreme
dry *summer*), 1821 (31·36 inches), 1824 (31·49 in.)
1828 (six inches in July), 1830, 1831. The dry
years, or those considerably below the mean, are
1802, 1803, 1807, (eighteen inches) 1808, 1813,
1815, 1820, 1822, 1825, 1826.

Thus we have (on the whole year and without
reference to a dry *summer*) *fourteen* years wet,
against *ten* years dry; and *eleven,* which are noted
for neither extreme.

Now, with respect to the four *seasons,* it was
found by a careful examination that their proportions
are nearly as follows:

	Mean Temp.	Inches.
Winter	37·20°	5·9
Spring	48·06	4·8
Summer	60·80	6·7
Autumn	49·13	7·5

It appears, then, that the rain is not at all in the
ratio of the Mean Temperatures : for Summer and
Winter exhibit almost like quantities of rain, though
differing so much in heat; while Spring is manifestly
dry, and Autumn wet, though nearly agreeing in
this respect. (*s*)

Again, the Rain from January to June is to that
from July to December, as ten to fifteen nearly :
but the rain from April to September, falls but about
half an inch (on the whole reckoning) short of that
from October to March; each period having thus a
fair moiety of the year. Thus we have with equal
quantities of heat very unequal amounts of rain;
and the reverse. But we will now proceed to the
Months, and shew where the difference takes place.

The figure which I now present to your notice
shews the Rain for each Month in a curve, with the
predominant winds for the Month at the top: the
whole on an average of thirty four years, from 1797
to 1830. The amounts are as follows, together with
the average number (in days and decimal parts,) *of
the days on which any rain fell*, in each Month.

Jan.	Rain, 1·90 in.	in days, 14·7
Feb.	—— 1·49	—— 14·9
Mar.	—— 1·39	—— 13·8
April	—— 1·84	—— 15·0
May,	—— 2·00	—— 14·5
June,	—— 1·94	—— 12·3
July,	—— 2·55	—— 14·4
Aug.	—— 2.15	—— 15·4
Sept.	—— 2·29	—— 13·8
Oct.	—— 2·41	—— 15·1
Nov.	—— 2·79	—— 15·0
Dec.	—— 2·38	—— 16·5

You perceive, 1. That the *amounts* of Rain,
beginning at a mean in January, go on decreasing

to *March*—which is the driest month of the year: that they then increase with some fluctuations to *November*, which is the wettest: after which the decrease begins again.

2. That in point of *frequency* of Rain, *June* is the driest month, and *December* the wettest; and that *March* and *September*, which differ so considerably in amount of Rain, are precisely alike in respect of the number of days on which any falls. (*t*)

The average wetness of *July*, which is manifest in the figure, is connected with the old tradition of Saint Swithin—that if it rain on that day, rain will continue to fall on forty days in succession. This is very seldom fulfilled, to the letter, *in any particular place*: but the fact is nevertheless certain, that about the middle of July in the majority of our summers a rainy period does commence, which continues till towards the end of August. It is a result of the great Solstitial disturbance attending the periodical rains between the Tropics, which by the time in question is propagated to the Latitude of these Islands: and these rains are accordingly always attended with manifestations of the Electric power.

Our *Autumnal* rains, though sometimes productive of thunder, arise from a different source: they are our own vapour condensed by the decline of our own Annual temperature, mixed with a large addition from the South, brought by the Equatorial currents. In November, (sometimes in October,) this process appears to be at its height—and this is, hence, our wettest season: and the latter six months are consequently by much the wetter half of the year. (*u*)

You will now be desirous of knowing how we measure the Rain and melted snows, and acquire this information. Here is a gauge, consisting of a bottle which is to be sunk in a small wooden case in the ground; and a funnel fitted to its mouth, in such a way as to exclude the entrance of any outside wet. The funnel has a turned brass opening, of precisely five inches diameter; and the glass cylinder, which is furnished along with it, is so graduated as to shew by simple inspection (when the water is poured in) the depth which has fallen on that circular area; and by consequence on the whole surface of the earth around. Two of these guages, kept several miles apart, will shew very nearly the same Monthly and Annual amounts of Rain. But if we go from the plain towards the hills, the quantity of rain is found to increase—and it is always much more considerable in stations on the *West* side, than in those on the *East*, of the great central ridge of hills which divides our island. While, at London, York and Edinburgh, the mean rain for the year differs but by an inch or two—on the West side, at the distance of fifty miles, the Annual amount is doubled. The *Atlantic* furnishes this excess; and the Easterly winds which dry one side contribute to wet the other. The real heavy rains however, on the East side of our island, come from the North sea, and always from an Easterly point. (*v*)

Though the amount of rain increases, as we ascend from the plain into a hill country, (both from the attraction of the clouds to the hills, and from its receiving the vapour from below in addition to that afforded by the district,) yet if we remove the gauge from the ground merely to a higher place in the atmosphere, (as to the top of a building), we find usually less rain collected in proportion as we ascend. The reason is, that in all misty precipitations the whole body of the vapourized atmosphere affords the water: thus what is formed into drops *below the level of the gauge* cannot be expected to be found in it.

When the Barometer falls with a Southerly wind, it often happens that the Thermometer rises; and *vice versa*, a Northerly wind, which raises the Barometer, lowers the temperature of the air. Hence, when the daily movements of the two are traced together on a scale, by passing a curve through the mean point for each day, they present many curious oppositions and coincidences. *Coincidences*, because there is yet a third condition, *belonging to seasons of rain;* causing the two to rise and fall together.

And the proportions in the rise and fall being frequently alike, (about two degrees to 0˙10 in. of the Barometer,) these curves present much symmetry; and are not only curious, but (in a sense) beautiful expositions of the results of Creative skill.

The three Curves in the figure represent portions of the 2nd, 8th and 12th, of thirteen periods of variation contained in the year 1807. Per. 2, is the course of the pressure and temperature during *frost*, in the first ten days of that year: Per. 8, the same during a *hot and dry* week in July of the same year: Per. 12, the same during *rain*, in the beginning of November of the same year. It will appear at once, by the relation of the dotted curve representing the daily Mean temperature, to the full one shewing the daily Mean of the Barometer, how they are meant to apply to the case. Each of them is laid down on the mean line of the Lunar period in which it is found. (*w*)

The subject of *Tides* in the Atmosphere was deferred, till we should have treated of the several appearances to which they belong: we may here bestow upon them what room we have to spare.

1. There is a daily flux and reflux connected with the Sun's altitude, and shewing itself like the tides of the sea twice a day. At *Calcutta*, in Lat. 22° North, where the smaller movements are not swallowed up in such extended sweeps of the curve as occur in these Latitudes, the Barometer was observed by Balfour, in 1794; to *rise* every day from six to ten A. M. and from six to ten P. M. and to *fall* during each intervening eight hours, thus making a regular daily tide. The like variation is to be traced in Registers of the Barometer kept further North, at such times as the air is in an extensive calm—or without any wind from North or South stirring. By careful observations however, made *hourly* and continued through many successive days, and then averaged on the whole number of days for each hour, on the Barometers of the Royal Society at Somerset House, it is now ascertained, beyond question, that we have the effects of the daily tide (though not its separate manifestations) here also.

In the midst of larger fluctuations of another order it appears, that from five A. M. to ten A. M. there is added to our atmosphere a weight equal to ·018 in. of mercury : that between the latter of these and five P. M. there is lost as much as ·026 in.—that an increase then follows between five and eleven P. M. equal to ·023 in. and a second decrease betwixt eleven P. M. and four A. M. of ·013 in.—thus making a regular daily tide as before. (*x*) Further observation will no doubt shew us further particulars of the difference, *in different seasons*, of these curious effects.

2. The more obvious *weekly tide*, (comparable to the neap and spring of the sea-tides) which is conspicuous in the curves of the Barometrical varia-

tion for these latitudes, has been more often observed
and speculated upon. There is a constant tendency
in the ordinary variation, to shew two elevations and
two depressions in each Lunar revolution, or in a
period of twenty eight days. These manifest changes
are, however, often set aside by a variation of a
different character,—more especially in seasons of
continued rain and storms. They are connected
both with the Moon's place in its orbit, and with its
declination North or South of the Equator. (*y*)

In the year 1798, a variation was observed to prevail connected with the Moon's *phases*, in such a
way as to present the elevations under the quarters,
and the depressions under Full and New Moon: and
this chiefly, during the part of the year free from
storms and continued rain. An account is given of
this year, with a plate of the curves, in Tilloch's
Magazine, Vol. vii. In the Climate of London the
subject is further prosecuted, and by the same observer: and it now appears that the *declination* is equally
worthy of attention in this respect with the *phase*.

The Moon's position, operating by the common
effects of the attraction of gravitation, influences
alike the course of the variable *winds*, the daily
variation of the *Temperature*, and the *Rain*, of every
year: but not in every year alike—there is a constant periodical variation of the variation itself.

The following are a few of the Results. 1. To
begin with 1807, the first year examined in the
work, the days on which a *Northerly* wind appears
under a Full Moon, (the spaces taken being weeks,
with the phase on the middle day), are double the
number of those that occur under the New Moon.
And the days on which a *South-west* wind blew,
under the New Moon, are to those under the Full as

thirty three to seventeen. The *South-east*, again, are six under New to four under Full Moon: while the East are eleven under Full to five under the New Moon.

2. The Rain of the year is found distributed accordingly: viz.

For the weeks under Last Quarter, 6·92 in.

For those under New Moon, 5·09 in.

For those under First Quarter, 6·17 in.

For those under Full Moon, 0·84 in.

The total for the Solar year being 19·02 in. we find that *not a twentieth part of the rain of the year fell in that quarter of the whole space, which occurred under the influence of the Moon at full.* (z)

3 Contrary to the state of the *Barometrical variation* in 1798, almost all the principal elevations of the Column appear in this year under the Full Moon, along with the Northerly winds.

4. Lastly, the *Mean temperature* of the weeks preceding New and Full Moon is lower, in this year, by two degrees, than that of the weeks preceding the quarters.

But we must not make the year 1807 a rule for others, any more than 1798. The year was remarkable (as has been shewn) for *dryness*, and that dryness under the aspect of the Full Moon: in 1808 the phase in question loses this character quite; and in 1816 (a very wet year) the Rain lies, two-thirds of it, on this side of the Moon's orbit, and chiefly in the week before the Full: the opposite phase being dry in proportion.

So much for the *phases*, but if we take it by the *declination*, the dry and the wet year agree in the distribution of their Rain: and this distribution, so far as it can be reduced to a common principle, appears to be as follows : *While the Moon is far South of the Equator, there falls but a moderate quantity of Rain in these latitudes : while she is crossing the Line towards us, our Rain increases ; and the greatest quantity falls, while she is in Full North declination—or most nearly vertical to us : but during her return to the South, the rain comes back to its lowest amount.*

The Mean Temperature, again, appears to increase along with the Rain, and to decrease as the quantity of that is reduced : agreeably to the known principle that heat is elicited by the condensation of vapour, and passes into the air. But if we have regard to the *phase*, instead of the declination as here, then the cold side of the Moon's orbit is also found to be the wet one: and *vice versa,* comparative warmth brought comparative dryness. And this obtains also as to the *years ;* the greatest depth of rain falling in those in which the Mean Temperature was lowest—and the warm years being dry, or at a mean in respect of rain. (*a a*)

It must be owned, here is matter in these state-ments better calculated to raise and stimulate than to gratify curiosity, respecting the principles of the Atmospheric variation. We have now, accumulated, a great mass of Meteorological observations, for the various latitudes and elevations above the sea, constituting all sorts of Climates ; and these acces-sible to all readers, in our Periodicals and in the Transactions of our Learned Societies, in Europe and abroad. Could gentlemen of the requisite Scientific knowledge, and who have plenty of leisure on their hands, employ both to greater advantage— the forward state of other sciences considered—than

in perfecting *Meteorology*? We should certainly
not be the worse off in our Agriculture, our Com-
merce, &c. for having more real skill in the art of
prognostics—we might derive considerable advantage from the study. At any rate, our time would be
innocently and agreeably employed ; and we should
find abundant occasion, in the course of our investi-
gations, to acknowledge the traces of HIS hand who
made all these things, according to his own unsearch-
able counsel, *in number, weight and measure*, for
our benefit :—and whose power sustains and brings
them forth in order, so that " not one faileth." Isa. xl.
12, 26. Let us then for the present conclude, and
silently adore!

LECTURE FIFTH.

The Clouds: their varieties or modifications of form and structure; the manner of their production, suspension, and resolution into invisible vapour, or descent in rain, snow, or hail: Thunderstorms and their effects: Whirlwinds, water-spouts.

❧

Thomson, in that fine poem of his 'The Castle of Indolence,' makes his 'man of special grave remark'—one of the idle, useless 'gentle tenants of the place'—a dreamer on the *Clouds:*

" Oft, as he travers'd the cœrulean field,
And mark'd the Clouds that drove before the wind ;
Ten thousand glorious systems would he build,
Ten thousand great ideas fill'd his mind ;
But with the clouds they fled, and left no trace behind !"

Industry however, combined with Science and applying its principles to their sportive and changeable forms, has at length rescued these ornaments of the sky and appendages to the gay landscape from their character of 'airy nothings,' and made them the subjects of grave theory and practical research. They are now shewn to be governed, in their production, suspension and destruction, by the same fixed Laws which pervade every other department of Nature. (*a*)

If air containing aqueous vapour in an invisible state, or (as a Chemical reasoner on the subject would say) in solution, be cooled below the *dew-point*, the water is separated in minute particles floating in the air; and forming mist, or diffused cloud. Again, if hot water be exposed to air of a lower temperature, it *steams*—that is, the vapour emitted from its surface is condensed in mixing with the air, the water thus produced appears in visible particles, and the constituent heat of the vapour passes into the air. This happens also with *ponds*, warmed by the sunshine of the previous day, about sunrise in our summer mornings; and even (in particular kinds of cold weather) with water newly pumped from a well. The small cloud formed in these instances commonly disappears presently, the general medium being too dry for it to be permanent —but in the atmosphere at large, and under the effect of the natural inequalities of temperature which obtain in large masses of air, the case is different. We see mists in the evenings of a fine autumn, and occasionally at other seasons, appearing suddenly in the valleys, whether near the water or not; which gradually fill these low places and even rise much higher; constituting a foggy atmosphere for the following day.

We have here the effect of the mixture, before described, of invisible vapour with cold air: the exhalation from the warm earth is condensed in a certain space—which by the evening *radiation*, (an effect I have already treated of in my second Lecture) has become suddenly colder than the air immediately above. The cloud then (if the mist continue and increase) fills this colder space; the portion of it which rises higher than that space being dissipated, as in the case of the partial formation of cloud before-mentioned. This collection of *visible natural steam*, resting on the earth below and cut off by a level sur-face above, so nearly resembles a sheet of water as to have been occasionally mistaken for an inundation, the occurrence of the previous night. (*b*)

Such is the origin and appearance of the *Stratus*, the lowest of the modifications of Cloud—the prin-ciple of which has been ascertained, by examining with the Thermometer the respective temperatures of the soil, or water affording the invisible vapour—of the cloud formed from it—and of the clear air above. (*c*)

The *Stratus* is then in its first formation an evening mist—but when permanent, and increased in depth so as to have risen above our heads, it con-stitutes *the fog of the morning*: and which has sometimes, as at the approach of a long frost, occupied the lower atmosphere day after day; to the great inconvenience of persons travelling on our roads, or passing in vessels along the coast. But the Sun we will suppose has broken through, and dissipated *this* obscurity and cleared the lower air—what then ensues? We see first on the blue sky some few spots, indicating the recommencement of the pro-cess of nubification there: we see these little collections of condensed vapour multiply and become *Clouds*—heaped as it were on a level base, and presenting their rounded forms upward: in which state they are carried along in the breeze, remaining distinct from each other in the sky. This is the *Cumulus*, or (as the name implies) the *heap*: which is the next modification that we distinguish in clouds.

By and bye, if there be a sufficient tendency to the process of nubification, and enough of vapour supplied from above, we see these heaps grow over their base, and assume somewhat of a mushroom or cauliflower shape;—presenting now a more level surface upwards, with drapery or festoons of cloud in their sides—connecting the original hemispherical form with the overspreading top. This is not, moreover, the only way in which the increase ensues; for, sometimes, the flat top is seen forming separately, and it becomes joined afterwards to the simple heap of cloud which before appeared—or they are even mixed irregularly, the flat forms and the heaps among each other; occupying the spaces every where till the sky becomes overcast, and presents the usual appearances of dense clouds. This is the *Cumulostratus*, or heaped and flat cloud ; it is not productive of rain, and it forms both in summer and winter the common scenery of a full sky.

If we examine minutely the higher region, especially after a clear time of some continuance, we shall perceive that it is often occupied by *threads* or *locks* and *feathers* of cloud, descending from above. These form what is called the *Cirrus*, (a Latin word denoting a lock of hair,) and they are capable of increasing, without any change in the mode of aggregation from the fibrous or linear, till they also fill the sky by themselves. More commonly however they are seen above the two former kinds ; which occupy the region beneath them, and float upon the clear air below.

The Cirrus, after having continued for some time, often passes to an intermediate form, which is called the *Cirrocumulus*. This is a system of smaller

rounded clouds; attached to each other or simply collected in a flat aggregate, and forming the mottled
or dappled sky; of which Bloomfield writes, so
justly as well as poetically, in his ' Farmer's Boy;'

For yet above these wafted clouds are seen
(In a remoter sky still more serene)
Others, detach'd in ranges through the air,
Spotless as snow, and countless as they're fair;
Scatter'd immensely wide from East to West,
The beauteous semblance of a flock at rest.
These, to the raptur'd mind, aloud proclaim
The mighty Shepherd's everlasting Name.

There is another intermediate modification, called
the *Cirrostratus*. It is distinguishable from both of
those just described: being more dense and continuous
in its structure—thick in the middle and extenuated
towards the edges. Overhead it is a mere bed of
haze, more or less impenetrable by the eye:—in
the horizon, and in profile, it often resembles shoals
of fish—but is liable also to put on the most ragged
and patchy appearances; making a very ugly sky.
And it is remarkable, and not a slight proof of
Creative skill, and if I may be allowed the expression,
of judgment in the picturesque, that all the fine-
weather clouds are beautiful, and those connected
with rain and wind mostly the reverse.

We have thus ascended from the mist on the
ground, (through six modifications,) to the very
highest appearances of condensation, in the Cirrus;
which may be seen from the tops of the loftiest
mountains, while the other forms are only skirting
their sides. We have now to contemplate the *Nimbus;*—or that form, in which the minute drops
constituting cloud, and remaining suspended as such
by virtue of their mutual repulsion, are by a change

in their Electrical state made to coalesce, and descend in drops of Rain.

The Nimbus, from a Latin word, used by Virgil himself in the sense in which it is here employed (*d*) *is a shower seen in profile*:—a body of cloud, having on its upper part a hairy crown, or brush of a Cirrous appearance, and connected below with a column of descending rain: it is, in other terms, a Cloud in the act of condensation into rain, hail or snow:—for it makes no material difference in the appearances, whichsoever of the three may constitute the product. In profile, and at a great distance in the horizon, it often resembles a lofty tower; raised by its greater height to a conspicuous place among the dark threatening clouds; and catching the Sun's last rays upon its broad summit and sides. In its nearer approach, it may always be known by being connected below with an obscurity, caused by the rain it lets fall, and which reaches down to the horizon. This is not the only mode in which rain may be produced out of the clouds:—but it is that in which we are best able to view the process, detached from the obscurity and confusion which fills the sky, when rain is forming all around.

Let us now, before we proceed to explain their nature, enumerate our *seven Modifications of Cloud:* they are, (*e*)

1. Cirrus; highest and lightest.
2. Cirrocumulus, ⎰
3. Cirrostratus, ⎱ intermediate.
4. Cumulus; detached hemispherical.
5. Cumulostratus; irregular heaped.
6. Nimbus; for Rain.
7. Stratus; mist or fog.

In proceeding to treat of these in succession, we

will begin with the *Cirrus*. I have already thrown
out some ideas on the process of Evaporation, and
the constant presence of *vapour* in the air ; which
may enable you to conceive of this source of cloud
and rain, as diffused through the whole depth of the
atmosphere. Were any proof of this needed, we
might find it in the large depositions of *snow* which
take place on the summits of the highest mountains ;
since the vapour affording this, must have previously
existed in the air at a still greater height. Vapour then being present, and condensation by
change of temperature admitted, we have only to
call in the aid of a third element, *the Electric fluid*,
to account for the peculiar appearances of the Cirrus cloud.

I have here a plate of glass, which I shall touch
with the knob of a charged Leyden phial, and then
project on its surface a fine resinous powder.—You
see that the powder adheres to the glass, *and is
collected in a sort of fibrous vegetation around the
parts touched with the knob.* Again, I shall place
on the prime conductor of the machine, this lock of
hair, and throw into it a slight charge. The lock
is not only set up by the charge, so as to make
every hair separate; but the hairs themselves follow
the motion of my hand, when carried over them—
*pointing towards any space into which they may
discharge the atmosphere of Electric fluid which
they hold.* (*f*)

These two experiments suffice to let us into a
view of the nature of the Cirrus. In a calm region,
or in air moving with a uniform gentle current, free
from the internal agitation which occurs below, the
minute drops of water produced by the condensation
of vapour are gradually attracted to a common centre;

from which the Electricity operates to great distances
to collect a cloud, in that curious manner in which
we see it form. And this cloud once formed is liable
to have its extremities directed, by the more powerful (though very diffused) charge of any neighbouring
tract, towards that space; in order to collect more
of the diffused water, and of the Electric fluid
which adheres to it.

The slow nature of this process, and the great
elevation at which it takes place, out of reach also
of the hourly fluctuations of the temperature below,
occasion the Cirrus to be at times a very persistent
cloud: it having been observed to retain its place,
with the same form, for two successive days, before
the collection of the watery particles could be com-
pleted. Hence also we see it at times stretching
across the whole sky from one horizon to the other;
indicating a very distant communication between
portions of the atmosphere differently electrified :—
as when one tract is covered with snow and the
other free from it; or when one is land, the other sea.

The pointing of the tails of Cirri upward is found
to indicate *rain ;* their downward direction is favourable to fair weather—and they are commonly seen
directed towards the quarter from which the wind is
about to blow. (g)

So much for the Cirrus: the next in order, for
height and lightness of structure—the *Cirrocumulus,*
by far the most beautiful of Clouds—may be accounted for by supposing the Cirrus fully saturated
with Electricity, and exercising no longer *the fibrous
attraction.* The common attraction of aggregation
now prevails, and the locks or tufts become rounded
masses, forming a system which still holds slightly
together; the several parts either joined or running

into each other—or the large occupying the middle
of the system, and the smaller disposed at intervals
proportioned to their diminished sizes, around.

This cloud must be considered as positively electrified; it is the attendant, and often also the certain
prognostic, of warmth and a vaporous air. Hence
should it appear during frost in winter, there is
every probability of a thaw by a Southerly wind,
propagated downward from the region in which it is
seen. It forms also commonly part of the preliminary
arrangements of a thunderstorm.

The *Cirrostratus* is yet further from the Electrical character of the Cirrus; it appears but weakly
charged, and as if placed between strata of air *of
different electricities*. It comes down and caps
the summits of lofty Cumulus clouds, when the
tendency to precipitation is considerable, but not so
as to produce rain; it gathers round these, as it
does upon the summits of detached and elevated
points of land in a mountainous country; and is
sometimes absorbed by the lower cloud in its further increase—sometimes itself increases, and forms
a tabular covering to the Cumulus; making of it
a very picturesque Cumulostratus. In the vast collections of clouds that precede thunder storms, after
great heat, the Cirrostratus may be seen flanking the
enormous Cumulostratus, and marking the probable
boundary between the positively electrified air
above, and the superinduced negative air below. It
appears also *as a low mist*, creeping by the mountain side up the valley, or resting on the higher
parts of the plain : but always distinguishable from
the *Stratus*—this being a dry Electric cloud with
particles repellent to other bodies—the Cirrostratus,
on the contrary, so moist, as to collect on the eye-

lashes as we ride through it, and run freely into the eyes. It is, lastly, the almost constant *accompaniment of seasons of wind and rain;* appearing both before the change, and serving then as a prognostic—and afterwards, during the restoration of fair weather, continuing for a while to linger in the sky. These two modifications, the Cirrocumulus and Cirrostratus, include the greater part of the *flat thin clouds,* so difficult precisely to describe, which fill our grey and mottled skies ; occupying a middle station between the elevated Cirrus and the denser heaps below; and by closer observation these are commonly to be traced, in their increase, to one or other of the structures here described.

The Cumulus and Cumulostratus are *dry* clouds —they must undergo certain changes before they can be resolved into rain : they disappear sometimes by evaporation, leaving a clear sky : sometimes evidently by mere dispersion, causing a general whitish turbidness. The Cumulus, being fed by vapour thrown up by the action of the sun, evaporates most commonly about sunset, when this supply ceases. The *Cumulostratus,* being supplied at once from above and from around it, is usually more persistent : it forms, in consequence of this arrangement, the proper medium of Electrical communication between the higher and lower atmosphere ; and when highly charged becomes at once *the true thunder-cloud.*

In the gathering of those tempests which succeed our longer periods of dry and hot weather, this cloud with its accompaniments forms banks in the horizon of peculiar magnificence, and attended with very singular appearances. At the commencement of the discharges to the earth, the high-

wrought forms of the lofty Cumulostratus are seen
to be replaced, here and there, by *the crown of the
Nimbus.* It is from the under part of this cloud
that the stroke descends ; and the whole body of
the Nimbus may be seen by an attentive observer
to be *kindled* at the time, like a burning coal, with
the Electric fire; the light of which is merely *re-
flected* from the surface of the adjoining clouds in
the other modifications. The discharges continuing,
and more of these passing successively to the state
of the Nimbus, the whole bank assumes by degrees
the common degraded appearance of clouds above
heavy rain. The *Electrical tension* was, therefore, the source of those wonderful natural
mouldings and carvings, on the walls of these
storied and buttressed äerial castles, which we saw
before.

The tension is now equally taken off below. Before the storm, we had heat and dryness, probably
with a slight breeze from the south-east: and the
lower air and surface of the ground were superinduced negatively, to that degree that the dust and
smoke filled the air, forming a dim haze, thicker on
the horizon than overhead ; in the depth of which
the Cumulostratus, the Cirrocumulus, the Cirrostratus, and (if the rain had already commenced
any where) the Nimbus, appeared in gloomy grandeur—motionless in their places, and the more
awful for being thus indistinct. But no sooner has
the Electricity found its way to the earth, taking
the readiest conductor it can find, than a very sensible
change ensues in the *air* ; the clouds get into brisk
motion with a westerly wind—and a clear sky, with a
refreshing coolness, (promoted now by the returning
evaporation,) succeeds to a sultry cloudy calm.

Thunderstorms differ at different times, in respect
of the *conductors chosen* by the electric fluid in its
descent. Those which occur in winter, and with a
N.W. wind, are deemed more dangerous to *buildings* than the summer storm : the clouds then fly
lower, and the leafless trees may tend (as good conductors) to prevent the *stroke* descending upon the
ground. In a storm which occurred Jan. 11, 1815,
in the Low Countries, no less than twelve steeples
were damaged; and several of the buildings on
which they stood set on fire: in this county of
York and the neighbouring one of Lincoln, we
have had two struck in a storm in the late winter.
(*h*)

Trees in full leaf and sap are most commonly the
objects struck in the open country: and the lightning, when it does not shiver the trunk, descends
upon the surface of the tree, ploughing up the bark
in a furrow of some width, often a part of the wood
along with it, to the ground. There is some reason
to think that a returning stroke goes at the same time
from the earth about the tree, upwards, as we may trace
at times two furrows—one more plainly connected
with the ground, the other originating about the
groin of the tree above, and descending. In either
case it is extremely dangerous to be under or near
the tree—as the human body presents a better conductor still: it is far better to remain in the open
field, and get well drenched with rain. Singular
as it may seem, *trees* do not die by the stroke, but
continue to grow on, unless shivered to pieces: the
animal on which it falls (as appears by the testimony of such as have been struck and survived)
neither sees, hears, nor feels any thing; but is instantly deprived of sense. The ancients, perhaps

from a notion of their being dismissed by the easiest possible death to the other world, accounted those struck by lightning the favourites of heaven. (*i*)

When a house is struck, the lightning takes a chimney, a gable end, or a corner of a hipped roof, and descends through the house, rending and tearing the imperfect conductors, and making what use it can of the perfect ones, to the ground. The wood-work and furniture suffer most, and the bell-wires are commonly all melted—and sometimes burnt to fluid glass, and dispered in most beautiful mossy figures on the wall. (*k*)

The wire in such cases has been seen to burn from end to end in a room, while the product of the combustion was dispersed on the floor in a fine black dust, which when swept up was found to consist of grains of black oxide of iron, resembling gun-powder. It would seem that a bell-wire of double strength would considerably add to the security of the house, (provided it were continued to the ground,) as the loops at the end have been found entire when the rest had been melted. (*l*)

The explosive effects of lightning in masses of masonry may be in part attributed to the sudden production of high pressure steam, from the water contained in the form of common moisture in the wall: but it sometimes goes directly through the most solid bodies, leaving a hole as if a shot had passed. It has been known to do this to a pile of several cast box-iron heaters, standing on a shelf, without melting the iron or displacing the separate pieces; the holes corresponding all through. In this case, as in those before-mentioned, we are obliged to infer *an instantaneous dispersion of the substance* by the wonderful energies of this all-pervading element of nature. (*m*)

Ships are peculiarly endangered by the approach of charged clouds, projecting as they do, high above the surface of the deep. Hence it is common to be provided with a chain conductor, to be run up, when a storm approaches, to the mast-head. The most curious effect of the stroke is, in these cases, to reverse the poles of the needle: so that when the bustle occasioned by the accident was over, and the vessel again under way, she has been observed by a consort to have been actually put about, and to be steering back towards the port from which she sailed ! (n)

The safest place for one timorous, when a heavy thunderstorm comes on, is in the basement of the house ; or on a feather-bed, or sitting in the middle of the room. Windows and fire-places should be shunned, in either situation; if a house be struck the windows are generally broken, but the glass is thrown *outward*. Fire-arms should never be left about at those times, primed and loaded : as the smallest spark of electricity in the lock will cause the charge to go off. (o) As to the actual danger, if you have heard the report you are safe from any fatal consequence of that stroke—but it sometimes takes the muscles, not the nerves, and wounds without putting an end to life. On the whole, the accidents to life by lightning are few, compared with those from other sudden dangers : and the most of these few by common caution might be escaped. (p)

The whirlwind and the waterspout are natural powers so nearly related in their origin, and mode of operating, that some persons have been led to consider them as the same phenomenon, exhibited in one case by land, in the other on the water. There is a plain distinction, however, to be made

K

between them :—the one presenting a more complex system of effects than the other; as will appear when we come to describe them in detail : we will take the simple *whirlwind* first. The miniature exhibitions of this occasionally dreadful spectacle are presented on our dusty roads, as often as hot and dry weather is about to break up for thunder. In a profound calm, too, with heat, we may observe the feeble precursors of the tempest, taking up from the ground to a great height leaves and straws, and other light bodies, to let them fall at a distance. VIRGIL notices this circumstance, as a sign of rain approaching:

> Sæpe levem paleam et frondes volitare caducas,
> Aut summâ nantes in aquâ colludere plumas. *Georgic* i.

When the dust is carried into this vortex, its form and movements are distinctly seen; and in the sandy deserts and arid dusty plains of hot climates the appearances, which whirlwinds occasionally put on, are even terrible to behold. (*q*) Our own are not always content it seems with the trifling plunder found on the surface : occasionally they are seen to take the cocks of hay from the meadow—the sheaves and hurdles from the field—the cloth from the bleach-ground—and the Russia mats and even the weighty greenhouse lights, from the garden of the nurseryman ; to disperse them over the country, or deposite them damaged in another place. (*r*)

If I pass a flat piece of wood rapidly through water, holding it upright, there is a sensible *running ripple*, which pursues the moving body through the fluid. A still larger whirling motion, and remarkably more persistent, ensues when a plug is taken out of the bottom of a basin or cistern, and the

water suffered to pass freely down through the
opening. The former of these experiments may
illustrate the horizontal, the latter the perpendicular
movement of whirlwinds. The air, heated by the
soil on which it rests, begins suddenly to ascend;
finding its way upward, as the water does downward, *in a spiral* : and having always contigu-
ous to it a further portion of heated air, the exchange
of the hot air for the cold is no sooner accomplished
in one place than it begins in another: thus the
effect is propagated laterally, as far as any great in-
equality of the temperatures may extend. Every
internal motion of a fluid requires, however, a motion in an opposite direction to counterbalance it,
and fill up the void it makes. Hence we have
probably *descending*, as well as ascending whirlwinds; at seasons when the atmosphere is greatly
disturbed in its density by local accumulations of
heat. (*s*)

But heat alone, we are sure, is not equal to the
whole of these effects. The most dreadful and
destructive sudden blasts and gyrations accompany
the movement of dense *clouds;* and the deposition
of rain and large hail from these is attended with
the strongest possible manifestations of electric
power. I shall place on the conductor of the machine a brass point made for the purpose, and
throw the electric fluid on the surface. It is now
impossible to draw a spark, as we did before the
point was attached. There proceeds instead, from
the point itself, a stream of invisible fluid, called
the electric aura, feeling sensibly cold to the hand;
like wind blowing through a keyhole. We have in
fact thus provided an artificial outlet to the artificial
electric atmosphere of the conductor; and if this

power, generated by art, and displaying itself at
the most on a few square feet of surface, be so
manifest to our sense, what may not be the energies
of a natural conductor, charged by the very circumstances of its formation, and presenting on its surface
more than as many square miles ? There is no room
to doubt, that the electrical energies are at times
suddenly emitted thus from a charged cloud, instead
of passing by explosion in a body of fire : and even
where we have the latter effect, there is also a portion of the natural aura in the shower. Hence the
strong electricity found in hard rain, hail and snow,
of which we shall treat hereafter. In all probability, the greater part of the sudden and almost
irresistible movements, (whether horizontal, or re-
volving, or ascending and descending,) into which
after hot sultry seasons the air is occasionally
thrown, are promoted by the same cause. (*t*)

At sea, these effects are found in the *Waterspout*
—a majestic and most powerful display of the exchange of opposite electricities, between highly
charged clouds and the superinduced surface of
the sea beneath them. The circumstances being
such as that the electric *spark* cannot be formed,
while the *tension* is nevertheless considerable, it
should seem that the atmosphere of the cloud
forms for itself *a discharging point*, out of the substance of the cloud—from which it begins to
descend in a rapid stream upon the water, throwing the air into a whirl as it passes through. The
surface of the sea where thus acted on appears
scooped out and forming a basin, around which the
negatively electrified water boils up in foam and
surge. The passage being opened, and the attractions continuing, the condensing cloud is at length

seen to pour down into the sea, *as it were through a spout;* the sea water at the same time mounting up into the cloud. The one in a direct way, in the inside of the column, the other in a spiral around the outside; until, the attractions being satisfied and the condensation fully effected, the spout is drawn up or breaks, and the whole ends in a very heavy shower. Such is probably the source of the appearances reported to us (with considerable variety in the description) by the many observing persons who have witnessed spouts at sea. (*u*) It is possible that the apparent mounting up of the sea water in a body, into the cloud, may be an illusion; and that the real effect may be *the filling upward* of the middle part of the whirl, with water actually condensed from the substance of the cloud, and brought together in that singular way. But that the whole is the result of the most powerful electrical action and reaction, between the cloud and the sea; and that the whirl in the air is but a secondary effect, like that caused by moving the stick through the water, is by far the most reasonable solution of the appearances presented by nature in this complex and wonderful operation. Spouts are dangerous to shipping, by the opposite impulses given to the sails and rigging as they pass through the whirl; in which the vessel has been found to heel over first one way and then the other, even when the spout has appeared to contain only cloud and wind, or a shower of large drops; but at times, when more fully charged, it brings upon the ship a deluge of fresh water from the cloud. And if dangerous to shipping (which at times fire guns, hoping by the concussion in the air to break the Electrical communication,) how much more

to small craft ? The natives of the South Sea
islands, though scarcely alarmed at thunder and
lightning, are accordingly greatly terrified by the
appearance of waterspouts. Throughout the Pacific,
says Ellis in his *Polynesia*, waterspouts of varied
form and size are among the most frequent of its
splendid phenomena. (*v*)   By the natives of the
Society Isles they are called ' *Huri, huri, tia
maona*,' that is ' turning, turning, rising upright
the deep': a truly descriptive epithet, answering in
brevity and energy to the reports of the mate, and
orders of the captain, in a case to which I refer in
my note. (*w*)' A waterspout off the weather bow, sir,
—making for us, sir !'—caught with all sail set.
' All hands on deck—let fly !'   Away then at once
go sheets and halyards :—and where the poor canoe,
with its square-sail and paddles, would have gone
to the bottom, the stately vessel escapes with a
drenching, and two or three reels of the masts!
' They that go down to the sea in ships—that do
business in great waters—These *see the works of
the Lord, and his wonders in the deep*.'   Ps. CVII.
23. 24. (*x*)

LECTURE SIXTH.

Electricity of the Atmosphere, and of Clouds—
especially of the Rain-cloud: sparks from
showers: Corposants: manner of the for-
mation of rain, hail and snow—the
avalanche—dew and hour-frost: the
Iris, corona, halo: the anthelion,
parhelion, paraselene: the
Rainbow and Meteoro-
logical glory.

The phenomena of thunder and lightning are sufficiently familiar to us. After the accumulation of clouds to a certain degree of density, and their approach towards the surface of the earth, there ensues a stroke between the two, of precisely the same character as the explosion of a charged jar or battery—though incomparably more loud and luminous. After one or more of these discharges there follows heavy and plentiful rain ; and in general a total change in the season. From extreme heat and dryness, the air passes to a moist, cool, and hazy state: and whereas, *before the storm*, we saw heavy clouds formed from day to day, which evaporated in the evenings without rain; *now* the atmosphere is become so impatient of the burthen, that it will scarce bear a thin sheet to pass over us without dripping on our heads. (*a*)

That *heat* is not the sole cause of thunder we may be assured, from the fact that snow-clouds at times explode freely—and that in a cold season. There must, indeed, be a condensation of vapour, in order to make a charged cloud:—and the reason why these no more accumulate and strike, when they arrive *after* thunder—is, that the air is now no longer an electric, but a conducting body. *In fair weather, the whole air is charged moderately with positive electricity.* We are able to collect this at any time, by insulating (that is by placing on its glass support) a pointed metallic rod, connected with plates of gold leaf, or with a pair of pith-balls; the divergence of these shews the charge—and their closing at the approach of excited, (that is rubbed and warmed,) sealing wax determines it as *positive.* A negative electricity, in which the balls close at the approach of excited *glass,* may be likewise observed at the coming on of a change of weather; or in the neighbourhood of clouds and showers. Either kind may be obtained from the air by means of a *kite,* flown in the common way, but insulated to some distance from the hand by a dry silk handkerchief, and having dependent from it to the ground a second string, consisting of thin *metal lace,* which conducts like a wire. Experiments of this nature are, however, attended with some danger, and should be conducted under the eye of experienced persons. The kite has sometimes shocked strongly, and might conduct a stroke: and Professor Richmann, observing at Petersburg with the insulated rod, received a ball of fire on the head, (on a sudden clap of thunder,) which killed him instantly. (*b*)

As the clear air, so likewise the *Stratus* cloud is
electrified *pos.*: and we have every reason to conclude that the *Cumulus* maintains its state and
magnitude, while it gathers above and evaporates
below, by virtue of a weak positive charge—by
which the repulsion of the particles is maintained.
The same may be said of the Cirrocumulus: the
Cirrus I have already described as a conductor,
and the Cumulostratus as *compound*. Our present
business is with those compound clouds, in order to
treat of their electric state, and of some particulars
which have not yet been mentioned, respecting the formation of the drops of rain and hail, and flakes of
snow. The Cumulostratus I have stated to be a
cloud, forming between a higher and lower region
of the air differently electrified—having probably
the opposite electricities weakly accumulated on its
upper and under surfaces. It is natural, in such
circumstances, to find on its flanks the *Cirrostratus*,
a plate of cloud nearly non-electric from this very
cause. When the superior positive atmosphere is
prepared to make of the clouds conductors to the
earth, the Cumulostratus pretty suddenly puts on
the character of the *Nimbus*, or rain-cloud—the
nature of which we have now to examine. (*c*)

Let us suppose an insulated conductor, consisting
of a pointed iron rod, mounted on a glass pillar
(under cover), and receiving on its upper length,
and on the inverted funnel through which this is
made to pass, whatever may fall from the clouds.
At the approach of a shower of hard rain or hail,
brought by a Nimbus moving with the wind, the
pith-balls open with negative electricity, and gradually close again : as the first drops or balls of hail
touch the conductor, they open *pos.*; and this

charge continues strong while the shower is passing over:—but the rain or hail gone by, the charge becomes again negative, and dies away in like gradual manner as before: lastly, there is left behind a slight positive charge. (d)

These facts prove that the central part of the space occupied by the rain-cloud is the focus of a strong positive electricity, concentrated by the diminution of surface in the water, as the drops come together and increase in bulk—that the circumference, again, is occupied by a negative charge, dependent on the former, preceding it upon its approach, and following it as it withdraws—the whole the result of a condensation, affording positive electricity, in the air above. These effects are perceived through a space of several miles in each instance; and they enable us to solve many singular appearances, attended with regular and irregular changes of the electricity, which persons who have given much time to this study have been able to observe. If we consider that, from the various forms of the base or ground-plot of clouds, and of the showers under them, the rod may be one while in the central area, another while in the circumferential space—one moment conducting a charge seated over and around it, and the next only influenced by a neighbouring one—it will appear that one simple principle may be applied here to a great variety of appearances: viz. that the electricity first evolved by the condensation of vapour is *positive*, becoming stronger (because limited to a smaller surface) as the drops become larger, and that this positive charge influences the surrounding air and clouds *negatively*, wherever it comes; and to a distance proportioned to its extent, and consequent intensity of power. (e)

During such examinations as the above, the rod is found to give sparks, at times, as freely as the charged conductor of the electrical machine—and even to strike to the neighbouring ball, placed for safety at two or three inches distance, and communicating by a stout wire with the ground. Messrs. Cavallo and Read, both practical Electricians, have left us many valuable observations on the Electricity of the Atmosphere, and of clouds and showers. It is, however, not from clouds, but from the rain, hail, and snow they afford, that the charge is commonly obtained.   Only five times in a hundred did *rain* fail to afford signs of electricity: snow is always positively charged—*hail* the same; and so strongly, that were we able to take sparks from every portion of it at once, the shower would more resemble fire than ice. It is the small rain that attends a close warm air, in which, at times, no charge can be found: hard rain affords it sometimes negative, sometimes positive: and there are many changes in a single shower. (*f*)

We need not wonder, with such effects from the approach of charged clouds and of showers, that persons have sometimes found themselves, and the horses on which they rode, electrified strongly by the air alone. In these cases, (of which we have a number on record,) the brim of the hat worn by the rider and the end of his whip-handle, with the horses mane and ears, present little luminous points, like those we see upon the comb of the rubber of the machine, and upon any point set upon the charged conductor. It has happened, further, in a highly charged snow-storm, that the trees and bushes by the road side have presented this appearance to travellers, at the end of every leaf and spray. In the case of a party of Genevese philosophers, who ascended Mont Breven at a time when thunder-clouds skirted the higher mountains on the opposite side of the valley, one of the party found himself attended by a kind of drumming noise, " bour-donnement," on the side of his head. This was found to belong to the gold button in his cocked hat—and those of the party who put it on their own heads found the same noise: they retreated, perhaps with necessary prudence, a little way down the hill, and the effect ceased. The button would no doubt have had on it a fine electric brush, had it been by night that this occurred. (g)

Still more considerable manifestations of the natural electricity occur upon metallic points, ex-posed to a thundry air. Pliny, the naturalist, takes notice of this, as happening to the pikes of a Roman legion—and modern bayonets have, I be-lieve, been found electric conductors, not inferior to the pike. The points and angles of crosses, &c. terminating public buildings, have been seen to exhibit very fine electric brushes—and these are, lastly, very frequent at sea, where they get the name of Corposants. The young sailor beholds with no small astonishment the mast head, or the ends of the yards on fire: he climbs fearlessly to the place, and tries this new gunpowder with his fingers—but he can neither burn himself with it (though it may run up the wet sleeve of his jacket) nor yet put it out—and the harmless visiter, whose presence announces (it is said) but a moderate de-gree of danger from the clouds, having staid its time, departs. (h)

Rain, I have stated, resulting from the aggrega-tion of the smaller watery particles of Clouds, must be more electric in proportion as the drops are larger, the same quantity of electric power being confined to a smaller surface. In *hail*, we have yet a more manifest reason for this—but it does not appear to make the whole of the case. It is easy to say of hail, that it is the drops of rain frozen in their descent: and this is found at times to be the case—a shower which encrusted the walls with ice, being caught on a sheet of paper, afforded at once drops of water, which froze where they fell; *shells* of ice, out of which the water had escaped by their breaking; and solid clear globules of *hail*. (*i*) But what shall we make of those larger hailstones, consisting of an opake nucleus thickly covered with clear ice—or of *bullets* of the latter, which, borne horizontally, and falling at a great distance from their source, break the windows, and disbranch the trees, and reap the crops and kill the game of a whole district—leaving the ground thickly covered with ice; or of those yet more terrible masses, *like round shot or fragments of stone*, which have attacked armies, killing in a few minutes thousands of men and horses—can these be the drops of rain merely frozen in the act of falling from the clouds? (*j*) It is very probable from circumstances, though it be beyond our power to scrutinize the operation as in some other cases, that here is what the Chemist calls ' a play of affinities' in the elements of the higher atmosphere: vapour parting with its whole electric charge and becoming water, and this water instantly concreting into small grains of ice, which as quickly run together into large aggregates, in a manner which, though we may faintly conceive of the operation, we should in vain strive to imitate by any process we know. Those little opake grains, like snow rolled up in pellets, which we see sprinkled on the ice in winter's mornings, having fallen in the night, are probably such concretes in the bosom of a cloud as, by falling through a vaporous air in summer, might collect ice enough upon them to make a regular hailstone. And we may admit the clear round balls, which do not, however, exceed a certain magnitude, to have been very large drops frozen aloft, and increased by collecting ice from the air below. With respect to long icicles, and pieces resembling ' fragments of a vast plate of ice' formed aloft and then broken, we must here leave them among Nature's wonders, and proceed to our next subject, the formation of *snow*.

The cristallization of water proceeds at an angle of 60 degrees : hence, next to the spicula, or little needle of ice, the next simple form is the little star of six rays, which, when it snows lightly in cold weather, we may detect falling on our clothes. On this may be built, by the addition of other needles at a like angle, and by the mixture in some instances of little grains of more solid ice, so great a variety of figures in the flakes of snow, that not less than ninety varieties of these, distinguishable by the use of the microscope in a cold medium, have been observed and figured by an ingenious person, filling a whole plate in the Philo. Trans., vol. xlix. (*k*)

It must be in consequence of the spicular cristallization, and of the constancy of an angle of 60° in the union of the needles, that snow is enabled to form those beautiful pensile drifts in which we behold it, and to lodge on the branches of trees and leaves of evergreens to so many times their thickness. Were the composition altogether of grains like salt, or even of spicula not endued with this peculiar

attraction, it would slide off as it falls: but by this
piece of nature's carpentry, (if we may so speak,)
it rests in large masses where it alights. One consequence of this accumulating property, the *avalanche*
is, however, inconvenient, and sometimes fatal (even
in our own climate) to man: and hence a just source
of dread to those who dwell at the foot of mountain
slopes, liable to be annually covered with new
snows. (*l*)

Dew is the product of a condensation of vapour
by the mere difference (in clear weather) between
the temperatures of day and night.   It is *propagated*
upward as the vapour rises and, with it, the cold
produced by radiation—which commences at the
surface of the earth : hence dew is found first at
the bottom of the valley, or near the stream : otherwise, once separated in those minute (and singly invisible) particles of which it consists, it *falls* in the
manner of the other products of vapour; subject
still to a variety of attractions differing in different
substances, which collect it in different quantities.
On this and the other phenomena of dew, much
instruction may be found in reading ' An Essay on
Dew, and several appearances connected with it,'
by Dr. Wells, published in 1814.   Dew is often
visible in the horizon, and to a considerable height
above it, as a purplish or reddish haze on the face
of the otherwise clear sky. (*m*)

Hoar-frost is of two kinds : one consisting of
spicula of ice, collected on the points and margins
of the leaves of grass; and on other bodies exposed
to the dew, with a like preference of the projecting
angular parts: the other is *granular*, and presents
the clear drops of the dew, frozen where they have
been formed.   *Rime* may be said to be a more

abundant and deeper hoar-frost—or, it may be
deemed the snow itself collected (not in the free
air but) on a solid body, which gives the first cristal its
support. It forms at times very magnificent scenery
on the trees, and on shrubs it is well worth examining with an eye-glass.   The freezing mist
borne by a slight breeze, attaches itself, like snow,
by virtue of the angular cristallization above-
mentioned, to the twigs of trees, the stubble, and
long grass, &c. in broad straps on the windward
side only—in such a way as to make a very singular appearance. (*n*)

Leaving the cold weather and its products, let us
now turn our attention to those luminous phenomena,
of which our own atmosphere and climate present
so great an abundance.   In fine autumnal mornings,
when the *dew* lies in large drops on the grass, if
we select one of the sparkling gems which it presents, we shall find that by varying the angle under
which it is seen, (moving for the purpose to right or
left,) we can draw out of it the prismatic colours,
blue, green, red, orange, yellow, in quick succession.   This is due to the refraction of light, in its
passage back from the posterior inner surface of the
clear globule to the eye.   We shall have occasion
to recur again to this subject in treating of the
rainbow.

Instead of the *eye* changing its place, let us now
imagine a surface nearly plane, like that of a large
cobweb on the bush, or of the collection of threads
left by the flying spiders on the herbage, called
gossamer—on this we may find the small drops of
the dew, ready placed at the requisite angles for
making a brilliant little *Iris*, exhibiting the prismatic colours as before-mentioned, but in circles.

The same effect ensues (or an analogous one, by
reflection after refraction,) when we see the moon's
disk through elevated haze, or a thin veil of the
Cirrostratus: it appears at times as it were in a
lanthorn, or globe of pale light,—this is called the
Corona ; at others the prismatic colours are finely
brought out in concentric rings around the planet.
These are called *Halos:* when near the moon they
are commonly more vivid in colour, and indicate
snow rather than rain : when distant from it, so as to
present a space of 40° (and even to appear oval,
from the ordinary atmospheric refraction affecting
the whole figure,) (*o*) they indicate wind and rain
within about thirty-six hours : sometimes, however,
a halo is colourless, presenting merely a distant
luminous circular band—which has been found, in
spring, to precede hot weather.

When the Cumulostratus has formed in various
quarters of a sky otherwise clear, and presents its
lofty sides, (which rival mountain cliffs in magnitude,) at different angles to a morning or afternoon
sun, we may perceive, again and again, on a part of
the cloud forming a recess in the middle region, a
space brighter than the surrounding portion, which
suddenly exhibits, by reflection, *an image of the
sun,* called the *Anthelion.* (*p*) The image is not,
indeed, always perfect, (though sometimes truly
circular,) and it is fugitive : but an attentive observer may satisfy himself at any time that it is
more than the ordinary reflection from the cloud.

The *Parhelion,* or common mock-sun, is an
analogous appearance, but more permanent ; commonly double, and seated at the intersections of
several luminous solar halos : for the sun, it should
have been observed, as well as the moon, has its

attendant luminous bands and circles, when it
shines through frozen vapours—floating aloft in the
air.   This is a phenomenon much more familiar to
the inhabitants of more northern climes than to
ourselves : its appearances are very various, and
philosophers have been more addicted to describing
and figuring these, than almost any other thing of
the kind.   We may content ourselves with a good
specimen, exhibited in this diagram, along with the
explanation given in the account of it by the observer. (*q*)

The *Paraselene*, or mock-moon, is produced on
precisely the same principles: both are connected
with the Cirrostratus cloud, and with the precipita-
tation of vapour at a great height—and give consequently an unfavourable prognostic to those who
desire only sunshine and fair weather.

The *Rainbow* may be found so fully discussed in
every book of the elements of natural philosophy,
that there will be the less occasion here to treat of
its principles ; but we may furnish what is more
wanted, an account of its relations and varieties,
and the necessary connexion of them with the
particular back ground on which they are painted,
thus superbly, by the hand of the ' Former of all
things,'—who was pleased to make his ' bow in
the cloud,' the token of assurance against a second
deluge, to mankind.   I may preface what we have
to say with one remark, that as in the former case
we had the mock-moon as well as the mock-sun, so
in this we have the *rainbow by night* (but commonly paler and less varied in its colours) when a
shower is seen receding against a bright moon. (*r*)

The rainbow, on a small scale, is seen in the
spray of cascades, (*s*) and is even producible at

pleasure, by the help of a good water-engine—the observer placing himself with his back to the sun. I have spoken of the refraction in the dew-drops, which we may change from colour to colour, by varying the position of the eye that views them: we may readily conceive of the same effect, in a shower of large drops, by the change of the relative position, or angle of vision, taking place not in the eye, but in the drops as they fall. So that each drop in its descent shall refract the colours in succession, from the red at top to the violet at bottom: but it is not with the natural prism as with the angular piece of glass, which we employ to throw the spectrum on the wall—the colours do not come out with uniform distinctness, and within the same spaces. In the bow itself, the red green and blue are conspicuous (hence ' the rainbow with three listed colours gay,' of Milton,)—the orange being lost in the red, the yellow in the green; and the darker indigo or violet, in the colour of the ground. In the complementary, or outer bow (which is only occasionally an appendage), we see the same colours, but reversed, the red undermost. I must now send you to Sir Isaac and his successors for the mathematical theory of the bow ; which would take up a full lecture of itself. (t)

The whole space included within the outer boundary of the upper bow, down to the horizon, is in reality affected by the refractive process. In a " most brilliant bow, which, together with a complementary one, was exhibited for about forty minutes, (in a thunderstorm in May, 1813,) the space included within the proper bow was very perceptibly *lighter*, and that without it, extending to the complementary arch, as much *darker*, than the rest of the cloud." Clim. Lond. ii. 203. See also this observation confirmed, p. 267. The effect described shews an obscure refraction through the whole space—but it is not always so obscure. In some showers of peculiar transparency, a succession of changes of the prismatic colours appears, bow within bow—shewing that, but for the interception of the rays by the rain itself, we should see the whole semicircular area thus occupied.

In very high situations, as on the Andes, the rainbow is commonly seen *circular*—there being no horizon to intercept the view of the lower part: circular bows are also seen momentarily on the spray of the waves, in a bright sun. On the other hand in a plain, when this phenomenon occurs with the sun setting in the West, it may be seen for several minutes after sunset: the lower part being first obscured by the shadow of the horizon, so that it gradually fades upward. (*u*)

When a rainbow is seen over water capable of reflecting the sun's rays, and thus mixing a new set of incident rays with the direct ones, the arch doubles in a peculiar way—two bows intersecting each other by a portion of the arch, and each appearing as if in its right place. (*v*)

A *white* arch is now and then seen in a mist, caused by the same *reflection* as the common bow ; but without the refraction, the drops not being of sufficient magnitude for the latter effect. And Lunar bows are, of course, though sometimes prismatic, much fainter. (*x*)

We will conclude this Lecture with the most *amusing* of all these phenomena, the *Meteorological Glory*. There are several accounts of this, (*y*) but we may take (as the most descriptive of the effect) the one observed by the author of the Climate of London, and described in vol. i. p. 224 of that work.

" On the 29th of the Seventh month, 1820, at Folkstone, Kent, the day was fine, with the Barometer at thirty inches, and the wind Easterly. There was a mist, of the kind which I commonly refer to the *Cirrostratus*, resting the whole forenoon on the cliffs towards Dover, and on the high land North of the town. Towards evening, the mist subsided from the cliffs, and appeared on the sea below them ; a body of cloud, which appeared to be *Cumulostratus*, showing itself also close to the horizon, on the high land above mentioned.

" About half-past six, p. m. walking with my family towards Sandgate, West of the town, we perceived that the mist on the sea was advancing and spreading itself Westward, and towards the shore ; and a body of it came at length close under the sandy cliff, on which we stood, at a height of about one hundred and forty feet from the sea. The mist was of various depths : a brig near the shore was at intervals completely hidden by it up to her topmasts : it exhibited a mixture of the *Cumulus* with the *Cirrostratus*. In this state of things, the sun shining clear above the Western horizon, our shadows were projected, together with that of the cliff's edge, upon the cloud beneath; on the surface of which, at the same time, each person could perceive, round the upper part of one of the shadows (which being distant were small, and rather indistinctly shaped) a luminous *corona*, surrounded by two faintly coloured *halos*. The outer halo was very large, compared with our shadows : it surrounded the whole group, and a considerable part of the circle was cut off by the shadow of the cliff. Consequently, when one of the party removed to a distance, his shadow was seen to pass the circle and appear by itself, without the *glory ;* notwithstanding which *he* continued to perceive the whole of the phenomenon for himself, around his own shadow ; those of the rest appearing to him at a distance, and also without it. We were able to continue these observations for about twenty minutes : until, the sun approaching the horizon, the shadows became too distant to be perceived, and the circles vanished. A thunder-storm followed these appearances, in the night of the 30th : after which we had again fine weather. The whole phenomenon was highly curious and interesting ; and the facility with which each of the party could either appropriate the *glory* to himself, or share it with the company present, suggested to me some reflections of a *moral* nature—in which, however, I shall not anticipate the reader."

We must not conclude the present Lecture, which has been occupied with the clouds and their products, without assigning to these natural structures, so admirably balanced in the sky, their proper rank *as a useful part of the Creation of God.* They pour upon us indeed, occasionally, the wasting flood, the fury of the blast, or the bolts of the tempest ; but even these bring relief from a state of sultry moist inaction, pregnant with disease and death. But they are, every day of their appearance, our benefactors in other respects. They shade us from the heat—they distribute and, as it were, economise the light, sending it by multiplied reflections into our dwellings—and with it the cheering influence it conveys. They warn us of the changes of the seasons—they announce the shower and bring it, letting fall the kindly blessing upon our labours ; and usually giving time to prepare against any hurt from its excess. They serve (there is no reason to doubt it,) even as blankets to keep us warm by night—checking the radiation from the soil ; by which, upon the sudden clearing of the sky, the temperature is known to descend many degrees, causing the most intense cold. (z)

I might add, and it may come not unsuitably last, (though not the lowest consideration,) that they are at times, lovely and glorious objects to behold ; inspiring cheerful sensations, and inviting the youthful imagination to revel in the changeable variety of their forms. Thus are we brought back to the enjoyment of the simple indolent admirer of nature, with whom we began ; but I trust this Lecture will be felt to have contributed, in some measure, to a more desirable, because more rational, view of the sky and its many changes—to a state of mind in which we may *read* as well as behold; and lay up stores for future *use* in reflection within ourselves, and in communication to others.

LECTURE SEVENTH.

Colour of the sky, Cyanometer : Night, twilight,
daylight : Shooting stars and fiery meteors :
Meteorolites, or stones falling through
the air : Refraction by the air, its
curious effects : Ignis fatuus,
so called : Aurora borealis :
Conclusion of the
course.

The subjects of the present Lecture (which con-
cludes the course) will be drawn from a variety of
phenomena, connected with light, electricity, and
magnetism, and found in all parts of the atmosphere,
from the probable height of its attenuated surface to
the air stagnant in marshy ground. We will begin
with *the blue colour of the sky.*

I told you in my first Lecture, that the air we
breathe *is a transparent fluid :* it is indeed emi-
nently so ; in the smaller portions through which
we behold objects near at hand. In the distant
landscape, the artist who studies its appearance,
discovers always more or less of what he calls the
air-tint—a colour varying according to the state of
the air, but most commonly, nay proverbially, *blue ;*

which distinguishes the distance, making hills appear lighter and more aërial in proportion as they are further off—and, hence, viewed with more *air* between. Our Saxon ancestors had a feeling of this distinction, for from these our old poets seem to have derived their term of the ' blea' for the country in view far off, and beyond the ' bourn.' Again, those who ascend high mountains (in the finest weather) find the distance always veiled in like manner, by a mistiness which the eye cannot fully penetrate—while they see clearly enough what lies beneath and around them. And from a greater elevation in his balloon, the best informed and most observing of our aëronauts, Baldwin, saw this veil (somewhat thickened) in so curious a way, that while all the distance was hidden, there appeared immediately under him, whereever he went, a portion of the landscape, as if painted on a circular space in the bottom of a huge bowl of porcelain, the brim rising to the height of the eye all around. (*a*)

But we will go deeper into the subject. Let my hearers then imagine themselves in the open air, in the latter part of a clear night in winter, the moon absent, and directing their view upward. What we experience in such circumstances, in beholding the intervals of space among the stars, the nebulæ and the milky way, may perhaps be most fitly denominated ' the blackness of darkness,' (*b*) a void from which no cheering ray returns. Let us now take our seats at a window, facing that part of the horizon where the sun is expected to appear: what we first perceive is a very faint light, breaking out as it were behind the earth : spreading fast upward, and to right and left. This *crepusculum* becomes presently a full twilight, pervading the whole sky : the *stars* going out as it advances, those of the first magnitude last; until at sunrise, the innumerable host of smaller luminaries is found to have vanished before the great orb of day. If the moon be present, there is cast over the azure ground, which sets off the planets and fixed stars, a degree of borrowed splendour proportioned to her fullness of phase:—but I shall not need to dwell on this, or on the reverse order of things, in the gradual appearance of the moon's brightness, and of the stars in succession, as the evening twilight declines. It is enough, for my purpose, that the sky by day *has on it brightness enough to eclipse every star*, and to make the moon resemble a map drawn in pale colours on a somewhat fuller ground of blue.

The source of this brightness (which is in itself of no mean service in affording us sky-light,) is the reflection from the atmosphere, and from the watery particles it contains. That it is chiefly, if not *wholly* due to this cause, is manifest from the great difference in the air-tint, as seen against a distant hill in fair settled weather *and before thunder*. In the former case, it is of some shade of blue or violet—in the latter it descends towards the indigo tint, and shows what we call a *lead* colour. But, viewing it as blue of various degrees of intensity, a Genevese philosopher, M. de Saussure, has furnished us with the means of observing and noting these degrees for use. The instrument represented in this figure, called a *Cyanometer* (or measure of blueness,) consists of fifty-two divisions round a circle, tinted with successive blues deepening from an almost white to the solid indigo. It shews, by being held up in such a manner as to be viewed along with the sky, to what number we may refer

the observation: the number increasing with the
deepness of the tint. About half the range of the
scale may be found in our own Northern skies:
those of Southern climes go far beyond these, as is
found by the surprise with which we behold, for
the first time, the lakes and rivers of the Continent,
the *natural* cyanometers of the country. At the
summits of mountains of great height, the dark
colour (from the absence of this reflection from above)
becomes so striking that the first guides who ascended Mont Blanc, catching the view of the sky
through the snowy peaks after their blind march up
the 'allée blanche,' were absolutely terrified at it,
and gave up the further prosecution of the enterprise
for the time. (*c*)

So much for the clear sky by night and by day—
now for what we see passing over it: and here also
we will take the night first. We can seldom be
abroad, during a few hours of the sun's absence,
without witnessing the phenomenon called *falling
or shooting stars*. Philosophers have attempted to
put these out of our own system into space, and
make of them considerable bodies: left out of the
planets, it should seem, at their formation, and
waiting to be picked up by our earth in its travels;
or even performing a cometary revolution about it,
and crossing its path in the heavens in the month
of November in greater numbers than at any other
time of the year! We need not go so far to account
for them, and even for larger meteors—these 'shooting
stars' are for the most part *electrical scintillations,*
drawn forth by the differing state of different
regions of the atmosphere: they may be seen to
descend on a group of thunder-clouds in the horizon, while the tempest is in full activity below—

and they have been found also to attend the aurora borealis. Their relation to *the differing states of the atmosphere below* is manifest: they appear before *wind*, and proceed towards the quarter it is about to blow from.

> " Sæpe etiam stellas vento impendente videbis
> Præcipites cœlo labi."—Virgil. Georgic. i.

Whence probably our Milton—

> " Swift as a shooting star
> In autumn thwarts the night, when vapours fired
> Impress the air, and shew the mariner
> From what point of the compass to beware
> Impetuous winds."—PAR. LOST, iv. 556.

In settled weather we see but little of these smaller meteors: and it is not likely if they were the indications of considerable (and these must indeed be considerable) masses of matter, kindled at a great distance from the earth, that they would shew so plain a relation to any sort of weather. Of the *larger fiery meteors* we have good observations, that suffice to determine their elevation: and this, as it appears, cannot be far from the surface of the aërial ocean that rests on our globe. I shall take first the description of the notable one of Aug. 18, 1783, seen by the author of this work (among, probably, many thousands besides,) as a large body of fire under various forms, which traversed Britain in all its length; and travelled, perhaps, in little more than a minute's time, from the latitude of Iceland to the Mediterranean sea.

By the various accounts of this meteor, inserted in the Philo. Trans. for 1784, we learn that it was first seen in Shetland, and at sea between Lewes and Fort William; that it appeared to persons at

Aberdeen, and Blair in Athol, ascending from the
Northward; and to an observer in Edinburgh as
rising like the planet Mars; that Gen. Murray,
F.R.S., then at Athol House, saw it pass over him
vertically—that it proceeded then a little West of
Perth, and probably a little East of Edinburgh,
continuing its progress over the South of Scotland,
and the Western parts of Northumberland and
Durham; proceeding almost through the middle of
Yorkshire, leaving York somewhat to the East, and
going S.S.E. That somewhere near the border of
that county southward, or over Lincolnshire, it de-
viated from its former straight course in consequence of a change in its appearance, comparable to
bursting: after which the compact cluster of smaller
meteors moved for some time almost S.E., thus
traversing Cambridgeshire, and perhaps the western
confines of Suffolk; but gradually recovering its
original direction, it proceeded over Essex and the
Straits of Dover, entering the Continent probably
not far from Dunkirk; where, as well as at Calais
and Ostend, it was thought to be vertical. That it
was afterwards seen at Brussels, at Paris, and at
Nuits in Burgundy, still holding on its course to the
Southward—and, lastly, it is intimated (though not
on any certain authority) that it was seen at Rome.
Thus it is thought to have traversed in all thirteen
or fourteen degrees of latitude, from N. to S. describing a track of 1000 miles, at least, over the surface of the earth; yet not visible to those who had
the best opportunity of seeing it in this island in
the middle of its course, (as to Mr. Herschell,
F.R.S. at Windsor,) for above forty-five seconds of
time.

As to its changes of form and attendant effects, it
was preceded in the N.W. (to an observer near

London,) by a glimmering light, resembling faint but quickly repeated flashes of lightning—the light increasing much, and then forming into a large luminous body, like electric fire, with a tinge of blue round its edges. Rising from the hazy part of the atmosphere, which might be about 8° above the horizon, it changed its size and figure continually, having all the appearances of successive kindling, and that not of a solid: it was sometimes round, at others oval, and oblong with its longest diameter in the line of motion: it was surrounded and accompanied in its whole course with a whitish mist or light vapour. This observer (Mr. Aubert, F.R.S. an astronomer,) then describes its leaving behind it several globules of various shapes, the first which detached itself being very small, the others gradually larger, until the last was nearly as large as the preceding body; soon after which they all extinguished gradually, like the bright stars of a sky-rocket—the light and magnitude greatest just before their separation. He estimated its apparent magnitude at two full moons—and says that the light was so great during its whole course that he could see every object distinctly. The altitude in the E. horizon 30° to 35° and the time 17 minutes past 9 p. m.

Lastly, Mr. Cavallo and other gentlemen, who saw it from the terrace at Windsor, determined the time to be seventeen minutes after nine (agreeing with the former observer),—the diameter of the burning body to be 1070 yards, and its height above the surface of the earth (confirmed very nearly in this by numerous other observations compared together,) *fifty-six miles.*

We have here the history of one of the most considerable of these moving masses of inflamed matter, which, at uncertain times, take their origin from the exhalations of the earth, and make their rapid transit from one region to another, *on the surface* (as it should seem) of our atmosphere; bursting often at the end, with a loud explosion heard below. That they are connected with its actual state, or with changes going on in it, we can no more doubt (from the accounts) than in the case of a thunder-storm: and between these two extremes, of the little white *falling star*, and the coloured *bolis*, or ball of fire having a sensible diameter, there are to be found on record accounts of meteors of every degree of magnitude, brightness, and length of passage—with not a little of variety in the form. (*d*)

But we have another strange thing to relate of them—so strange, that when it was attempted to be brought again before the public, (not many years since, and after a long period of neglect,) the history of *stones fallen from the sky* was treated by very many as a deception of sense on the part of the relators; or, as a wilful imposition on the credulous. Certain it is, however, and ascertained on the most satisfactory evidence, that some of these bodies have let fall not merely single stones, but a whole shower of these, on the tracts over which they have passed: and the composition of such bodies, collected from various countries, as found by chemical analysis, has confirmed the opinion that they have all one common origin. Now, though some philosophers have preferred to make, even of the smaller meteors, very distant bodies, extraneous to our atmosphere; and have brought the larger from the *moon* in the form of projectiles, there is very strong ground to presume that the white trains which falling-stars leave behind them, (at some seasons and not at

others), are in reality the *ashes* of the combustion
of a mixed inflammable substance, chiefly gaseous;
which residue, at first incandescent, gradually van-
ishes by cooling where it is left, and subsides un-
perceived through the air.   Again, that the solid
stones, which have the appearance of earthy con-
cretions, suddenly formed in a heated medium and
glazed by flame on the surface, are the cinders and
slag of a still larger aërial fire; of a *furnace*, indeed,
(for such was the aspect of the meteor of 1783),
carried through the air with the apparent falling
of burning coals from it, which become extinct in
their descent.

The composition of these *Meteorolites*, their history, and probable origin in vapours, carried up
from the earth and fired aloft, would form of themselves a very interesting lecture, but which does not
enter into the plan of my course.   Suffice it to say,
that though the stones in question have been ascertained to have fallen from the heavens by day, in
various parts of the world, and from remote an-
tiquity—it was not so easy (and yet modern observation has been found equal to this,) to prove their
connexion with the *bolis*, moving through the air,
and drawing a flame behind it.   This, however,
being proved, with the fact that hydrogen gas is
capable of dissolving various bodies, even iron: and
that it is naturally evolved, mixed with carbon in
the gaseous state, in very large quantities (even
from every piece of stagnant water, in the autumnal season,) we have a right to presume that, on
occasion, it is collected in vast fields, to be fired by
electric explosions, or by some play of affinities in
nature, of which we have not as yet a proper con-
ception; and (the gases burning out) to let fall the

earthy and metallic contents, precipitated and agglu-
tinated as we find them in the aërolite. A meteor
of such a nature, covering an extent of many acres
in the atmosphere (as these do), may very well
afford a brilliant light, and (though but slightly
charged in proportion to the mass with solid matter)
also the residue of the combustion, which descends.
But a body of the size of the largest aërolite, coming
solid from space into our atmosphere (admitting it to
take fire on first entering,) would form but an in-
conspicuous object in its descent. (*e*) We must
now proceed from this interesting inquiry to the
remaining subjects of the Lecture.

The *Ignis fatuus* (of the philosopher,) or Jack-o-
lantern (of the rustic,) might be deemed a meteor,
generated at the surface of the water affording the
carbonated hydrogen gas, of which I have been
speaking, were it not that there is some doubt
whether it be meteoric, or ignited at all. It should
seem from what we meet with in old books, and
from the proverbial aspect of the thing in its denomination, that this appearance was formerly
very common in England: it is now (perhaps from
our improved drainage) very rare. I must refer
the curious to the descriptions given of it by those
who have seen it, with this cautionary remark—that
it is quite as easy to apply the most sensible and
minute of these accounts of a supposed meteor, to a
swarm of insects, of some species not hitherto noticed *as luminous in the dark,*—though possibly
classed by entomologists as they appear in daylight: but which, from the lessening of their breed-
ing haunts, are now seldom, if at all, observed here.
(*f*)
The grandest of all the exhibitions presented to

us in the sky, when seen in its full perfection, is perhaps the *Aurora Borealis*. To behold the heavens in flames in their whole breadth, and up to the zenith,—to see spires of luminous matter tinged with the colours of the rainbow, shooting from a vast arch of light, thrown as it were over a dark space in the north ; and after reaching their greatest altitude, chasing one another with the rapidity of lightning through its whole extent,—then as quickly subsiding, blending and changing in a thousand ways, and at length settling in a still milky whiteness—this is a spectacle which must be viewed and contemplated for some time, to enable us to recall it all in a brief description. It is no wonder that the antients, terrified at its appearance, and mixing largely the products of imagination with the realities of nature, should have recorded it on many occasions as a prodigy—as the combats of celestial hosts, exhibited thus to the view of mortals, and forboding like dreadful conflicts among the powers on earth ! (*g*)

All this display is believed, however, to result from the Electricity excited and liberated in one region of the atmosphere, and passing off to another, where it finds a readier descent to the earth. Before I proceed to some experiments illustrative of the subject, we may properly go a little into the history of these phenomena. The Aurora is a common appearance in the winter of the Polar circle. There it has been seen, by some navigators to the North, by others to the *South* of their position ; highly luminous, and at times even audible in its movements, (so they report,) making a hissing and crackling noise like that of a flag flapping in the wind.

The Aurora Borealis is only an occasional visitant
in these latitudes, but usually seen several times in
succession when it appears. It was thus named by
Gassendi, in 1621: who then observed and described
it in his ' Physics,' after it had for about forty years
escaped notice; it seems to have appeared but rarely
in the whole 17th century Like the more powerful
manifestations of the Electric energies, it is found
when it approaches the zenith to disturb the magnetic needle, causing it to vary and fluctuate to
the East and West of North. It differs here in form,
intensity, colour, extent, and duration; and the
reports of our officers, engaged in expeditions to the
Polar Sea and regions within the Artic Circle, shew
it to have the like variety there. (h) In the Philo.
Trans., abridged, vi. 213, is a paper by Dr. Halley,
giving a particular description of an Aurora, the
latter part of which he had witnessed on March 6,
1716. He says (after expressing regret at not
having seen the whole with his own eyes, as he had
all the several sorts of meteors he had read of,)
" This was the only [meteor] I had not as yet seen,
and of which I began to despair; since it is certain
that it has not happened in this part of England,
to any remarkable degree, since I was born [1656] :
nor is the like recorded in the English Annals since
the year of our Lord 1574—above 140 years since, in
the reign of Queen Elizabeth. Then, as we are
told by the historians of those times, Camden and
Stow, eye-witnesses of sufficient credit, for two
nights successively, viz. on the 14th and 15th Nov.
that year, much the same wonderful phenomena
were seen, with almost all the same circumstances,
as now." He refers to other accounts of appearances of the Aurora, viz., at London in 1560 and

1564: in Brabant, twice in 1575; in Wirtemberg,
seven times, in 1580 and 81; in 1621, all over
France; in 1707, of short continuance in Ireland,
at Berlin and Copenhagen,—and lastly in Britain,
five times in the space of 18 months, in 1707 and 8.
I shall not need to say more, to show that the present generation of the curious in natural phenomena
may felicitate themselves in seeing, as well as hearing so much of it; and I may now introduce the
particulars of three observations of Auroras of old
date, not surpassed in their singular appearances by
any of modern times.   1. At London, Nov. 10,
1719, by Dr. Halley, Sec. R.S.   About five in the
morning, an entire canopy of white striæ *seeming to
descend* from a white circle of faint clouds, about
seven or eight degrees in diameter: which circle
would vanish on a sudden and as suddenly be renewed: the centre not precisely in the zenith, but
rather 14° to the South of it.   None of the striæ
came lower than to about 30 or 40 degrees of altitude, and they *seemed not to have ascended* from the
horizon.   In the night following, a strange stream-
ing of lights was seen in the air, from half-past nine
to eleven; when a fog came on and obscured them.
During that whole time *there ascended* out of the
E.N.E. and N.E. a continued succession of whitish
striæ, *arising from below*, which, after changing
into a sort of luminous smoke, passed overhead
with an incredible swiftness, not inferior to that of
lightning: and, as it passed, seemed as it were
gilded, or rather as if illuminated by a blaze of fire
below.   Some of the striæ would begin high in the
air; and a whole set of them, subordinate to each
other, like organ pipes, would present themselves
with greater rapidity than if a curtain had been

drawn from before them; some of which would die
away where they first appeared, and others change
into a luminous smoke, and pass on to the Westward with an immense swiftness. Philo. Trans. abr.
vi. 441. Here we have a descent, as well as an
ascent and horizontal motion of the beams, and a
canopy near the zenith—of which more in the two
next accounts.

2. Rev. W. Derham, F.R.S., describes an Aurora
observed Oct. 8, 1726 It began about 8 p. m. opening and shutting as if a curtain had been drawn and
undrawn before the streamers, which soon extended
to every point of the compass, unlike those of 1715-16.
They were mostly pointed, appearing as flaming
spires or pyramids—some truncated and reaching
but half way up—some had their points reaching up
to the zenith, or near it, where they *formed a sort of
canopy*, or thin cloud, sometimes red, and sometimes brownish; sometimes blazing as if on fire, and
sometimes emitting streams all around it. This
canopy was manifestly formed by the matter carried
up—which seemed to ascend with a force as if im-
pelled by some explosive agent below; like that of
March, 1715-16. This forcible ascent of the stream-
ing matter gave a motion to the canopy; sometimes
a gyration like that of a whirlwind, which was
manifestly caused by the stream striking the outer
part of the canopy: but if it struck the canopy in
the middle, then all was in confusion. In some
part of the time the vapours between the spires were
of a blood-red colour, which gave to those parts of
the atmosphere the appearance of blazing lances
and bloody coloured pillars! In the North and
South the streams were perpendicular to the horizon; in the intermediate points, they inclined towards the meridian. Idem., vii. 181.

N

3. On Jan. 5, 1727, 7 p.m. at Liverpool. From the northern parts arose several streams of light, as if from behind a black cloud: (*i*) they were innumerable, and shot up to the zenith, with a motion not to be followed by the eye. They had also another motion, which seemed to be sideways; their higher ends terminating sometimes in a sharp point, sometimes in two or three. They appeared from N.W. to N.E., but were brightest in the North: their colour was pale, like that of Jupiter through a telescope, but not so bright. Most of them reached the zenith, where, mixing with each other, they *whisked round and formed an appearance like the curling flame of a glass-house fire.* They had a very irregular motion, some turning inwards, some outwards, like the pendulum-spring of a watch. This circular light was the brightest; and seemed to occupy near 10° of the highest part of the hemisphere. Several strokes of light seemed to dart from it to the South, but died before they got any considerable distance. About ten o'clock, the *whirling light in the zenith* appeared of several colours; as blue, green, yellow, and reddish [orange?]. Idem., p. 194.

Making every allowance for the imagination of the spectator in the former case, placing as it were a solid dome over the space around the zenith, for the lights *to break against,* we have here concurring evidence of the ascent in a spiral of the streams, *upward from the earth*: the perspective would bring them together near the zenith, to every one's eye who saw them, in whatsoever place. But we have a third and very curious fact to add: old Stowe says in his *Annals,* in 1574, on Nov. 14, " were seen in the air strange impressions of fire and smoke to proceed forth of a black cloud in the North towards the South. The next night following, the heavens from all parts did seem to burn marvellous raging; and over our heads the flames from the horizon, round about rising, did meet, and there double and roll one in another, as if it had been in a clear furnace." Idem., p. 184.

That we may not make a Lecture of this particular phenomenon, we must here, however, break off these descriptions; referring to modern publications for a great variety of accounts besides. (j) But it is pretty clear, from the appearances put together, that we here behold the free Electric fluid in prodigious quantities, making its passage from a part of the earth where it is repelled by a surface covered with snow and ice, to another where it may descend with freedom. The particular form and direction of the columns in which it mounts or descends, or of the natural conductors which it uses in those elevated regions, constitute the most curious part of the subject. For it appears, by the most minute and exact observations of those competent by their mathematical and physical knowledge, to decide the question, that the Aurora is in effect an Electromagnetical process of nature, and governed by the following principles. 1. The spires or beams, however diversified in their appearance by optical deception, are cylindrical and parallel to each other, at least over a moderate extent of country. 2. These cylinders of electric light, or the conductors through which this moves, are all magnetic; and parallel to the dipping needle at the places over which they appear. 3. The height of their superior part (where the Electric fluid appears to take a horizontal direction, moving away in rainbow arches,) is about 150 miles above the earth's surface. 4. The beams (however varied in appearance by the effect of the perspective in the sky,) are similar and equal in their real dimensions to each other. 5. The distance of the beams from the earth (at bottom,) is equal to their length, nearly. 6. That appearance which we call the horizontal (or still) light, and which is always situate near the horizon, is nothing but the blended lights of a group of beams or flashes; which constitutes a large luminous zone.

Such are the propositions advanced by Dr. Dalton, as far back as the year 1793, and republished in 1834, in the second edition of his Meteorological Observations and Essays. He thinks the beams may be 75 miles in length, and one-tenth of this or $7\frac{1}{2}$ miles in diameter; and that the conducting medium is *iron*, in a state of division or diffusion approaching to the nature of elastic fluids; consequently, that the whole system is subject to the earth's magnetism: [and for aught we know, those magnetic columns may be present every where for a great space around each magnetic pole; being only visible when lighted up by the passage of the Electric fluid.]

But when we have stated this beautiful and well-conceived magnetical hypothesis, (built on actual observation), we must recollect that it has still to do with Electricity; and that the moment the luminous matter is out of those trammels we must expect to see it as in the description given, flitting away like clouds, or bounding like billows, or flashing with the rapidity of lightning to the place of its destination; and this being through a space 150 miles above our heads, we need not wonder that on some occasions it has been seen at the same time in many of the principal countries of Europe, far distant from this island; nor (if we consider that the magnetism of the earth itself has its periods, from the changing of the places of the magnetic points), that the aurora should at times be seen often in the temperate zones, and then for a long space be confined in the vicinity of the arctic and antarctic circles,—for it has been observed as well near the south as near the north pole of the earth. (*k*)

We may now conclude the course with a few reflections. A knowledge of the phenomena of nature, *and of their causes,* confers on the civilized and instructed a prodigious superiority over the savage and uncultivated portion of mankind. The antient poets and philosophers were sensible of this; and we meet with passages in their writings which show their contempt for the credulity of the multitude, and the happy state of those who were raised by a better education above their slavish fears. But the possession of this good, the fruit of knowledge, required still the safeguard of a good conscience,—of integrity within. And, thus provided with the antidote to the bane of ignorance, who would not desire and endeavour on all proper occasions to impart it to his fellow-men? The Christian religion enjoins this: benevolence of heart will ever prompt to it. What can we expect from the degraded by superstition, from the embruted by moral and mental neglect, from the *sunk* in error, but a conduct suited to their lower nature—a spirit befitting the gloomy region in which they dwell?

The ancients were grossly ignorant of the laws by which the universe is governed. They beheld in the tail of every comet the presage of 'pestilence and war,' and saw armies engaged in mutual slaughter, and the very knights pricking their aëry steeds to the combat, in that fine and harmless display of the electrical and magnetical energies which we have just now described and explained. Need we wonder, then, that they were easily persuaded by interested priests to take natural objects first for the symbols, and then for the direct objects of their worship? That the sun and planets, the earth, the air, the ocean—nay, the very stocks and stones which themselves had shaped into beauty or deformity came at length to receive Divine honours, (*l*) and each remarkable tree or river to have ascribed to it a presiding deity within?

In the midst of this darkness it pleased God to preserve in one race of men, the Hebrews, the knowledge of His own Being and attributes. Among them was found that great primæval truth, ' In the beginning GOD created the heavens and the earth,'—a truth with which *we* are made, from infancy, familiar; scarce conscious of the time when we possessed it not! Provided thus by Revelation with His gracious gift of knowledge, (for it was freely given, not taken by the force of our spirits from the fountain of wisdom and might), *we* may contemplate Creation with other eyes ; and having walked forth among the monuments of his power, and seen a portion of his wonders, be ready to say, with Milton—

> " These are thy glorious works, parent of good,
> Almighty, thine this universal frame
> Thus wondrous fair!   Thyself how wondrous then,
> Unspeakable : who sits above the heavens,
> To us invisible, or dimly seen
> In these thy lowest works : yet these declare
> Thy goodness beyond thought, and power Divine !"

THE END.

APPENDIX.

NOTES: LECTURE FIRST.

Note *a*. *Labor omnia vincit improbus.* In consonance with this maxim, the young Lecturer may be reminded of the necessity of *taking pains* to please and instruct his audience: he will find his labour lessen, in proportion as he gains *experience*. But even with this help, he must not trust himself before the public, without having first gone over the whole Lecture with his assistant (for an assistant will be found useful in most of the Lectures;—quite needful in the *First*, and some others) in order to see that every thing is in order, and in its place. Such a degree of attention will save him much delay; perhaps too, some mortifying accidents and failures.

b. It is obvious that we cannot have *direct* proof of the existence of a definite surface to our Atmosphere, at this or any other height: but the statement given is in agreement with the opinion of the late Dr. Wollaston; and of other eminent philosophers.

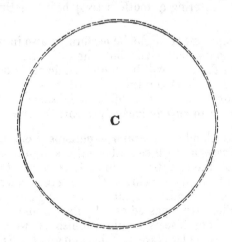

c. The representation of these proportions as in fig. *c*, will require a circle (on white paper) of two feet diameter. Out-

a

side this, draw a lighter circle, 0·24in. or *within* a quarter of an inch, distant from it. The space between the two may be tinted with light blue, and a Portfolio should be provided of the requisite size conveniently to hold this along with other Figures, to be hereafter mentioned.

d. It may not be convenient to every Lecturer to have with him the standing scales, and other apparatus necessary for this experiment; which however may be had at the shops: but—

e. The simpler method may suffice of a small balance, carrying at one end *a cork*, at the other a brass weight, in counterpoise with the cork *in a dry state.* This will prove, hydrostatically, that the cork displaces a certain weight of air, and is buoyed up by it: for the fluid medium being removed, the cork is seen to preponderate.

f. Air is poured into the furnaces of foundries and smelting works, from iron cylinders of large diameter; in which solid pistons rise and descend, in the manner of those in the Air-pump. The effects of the stream of air on the fuel are very powerful; and the noise almost deafening.

g. Care should be taken to bring the racks and pinions to full contact (entering a tooth or two) before setting them in play.

h. The experiment might be performed also in water, in a glass jar of convenient size : but the difference is by far less striking, and the time will be found sufficiently occupied by repeating it in air—as the audience will most probably desire to have it done—after the exhibition *in vacuo.* The gauge should be down to *half an inch* of pressure for the experiment.

i. k. The Condor (or *contor,*—perhaps from its circling movements) ascends with ease above the summits of the lofty mountains of the country in which it is found, to obtain a wider view of the plain below. Carrier pigeons fly at a rate that must exceed a mile in a minute ; through great distances in a day. The spaces passed over by the common swallow, in its daily pursuit of insects, would astonish us, were they computed from actual observation. Flies and gnats are remarkable for their power of accompanying the horses attached to a carriage ; and seem to be themselves in some way attached to the horse—their many evolutions about him notwithstanding.

l. Franklin made very successful experiments on hearing, *under water*, distant sounds produced in the water. See his Works vol. ii. p. 335. Ed. 1806.

m. The supporters of combustion exclusively of air must be familiar to students in Chemistry; but they are not at present to our subject. It will be proper to accompany this part of the subject with a wax taper, burning *out* under a bell glass; to shew the speedy consumption of the oxygen present.

n. The early experiments with the air-pump included these on animal life; examples of which may be found in books on the subject of Pneumatics. The spiracula of insects are not liable to the like injury in a vacuum, with that which disables the lungs of birds and quadrupeds. A large *libellula*, (or dragon-fly) being placed under a roomy glass on the air-pump, and a good vacuum made, became soon incapable of flight, attaching itself closely to the plate, at its junction with the glass. Here it remained motionless till next day, the vacuum being still maintained; when on the readmission of the air it speedily revived, and flew about as before!

o. The membrane should be of the kind made of intestine, called gut-skin. The thinnest *bladder* is found too strong for this use. It will be proper to take off the little *siphon-guage* in this experiment; as it is subject to fracture by the sudden recoil of the quicksilver.

p. The *squares* (as they are termed) sold for this purpose should be chosen by their *lightness*, and carefully kept, several in a case properly divided: the brass cap with the valve should be transferred to a new glass, and secured with cement, *on the conclusion of the course.* It is usual to have a brass wire guard over the bottle: but it is more needful to see that the fragments of the glass do not get into the pump.

q. This experiment answers best with the *medlar*, I have likewise found *grapes*, half dried, and *apples* kept to a certain point, succeed very well: but it is almost too *minute* for an audience; unless the fruit be operated upon in a small receiver on a transfer-plate with a cock, and so handed about the room.

r. Cold-drawn linseed oil, froths under an exhausted receiver, like so much beer. See " Climate of London :" *Intr.* xi.

s. The apparatus for this very striking experiment may be had at the shops: the pump should be filled some time beforehand, and the action of the valves tried. The Lecturer will perceive (on reflection and trial) that the same glass receiver may be applied to a variety of purposes: *this* operation requires one about a foot in height, five inches in diameter in the body, and two in the neck: which will be found the largest he needs to take with him for the course.

t. A common stiff spiral spring of wire fixed on a stand, so as to admit of being pressed down by the hand and making its recoil, exemplifies pretty well this first view of the subject: a more delicate spring, required to shew the *balance of pressure and elasticity*, will be spoken of further on. This spring will require a glass cylinder for lateral support.

u. It would surely be possible to vary the construction of this apparatus, in such a way as to render it unnecessary to carry from place to place *a mere weight of fourteen pounds:* procurable at any grocer's shop for the evening's use! The wooden block, placed immediately on the bladder and filling nearly the diameter of the glass, would sufficiently steady it.

v. There is provided in the shops a close tube mounted on brass, to cover the open one, by connexion with the receiver used in *s*,—or with a smaller one.

w. The Lecturer may take this opportunity of pointing out on the machine itself (or with the help of a drawing) the several openings, chambers and valves, with the manner of the operation of the pistons: *which see*, described in works on Natural philosophy and Pneumatics.

x. The instrument mentioned in Note *v*, for raising a column of quicksilver by the expansion of the air, will serve this purpose; a *bead* only, or short column, of quicksilver being now introduced into the tube at the lower end; which is to be left open in the empty bottle, the tube being screwed in.

y. This experiment should be performed under the largest of the glass receivers, by means of a rod with a small hook on the end passed through the collar of leathers in the brass cover, and another hook screwed into the under side of the cover; the thread should be adjusted as to length, &c. beforehand, and should not be too thick: the first trial of the sound should be made *under the recciver, with air in it*—then the trial *in vacuo.* As to the experiment with the pin on the piece of timber, it is one precisely analagous to those commonly made in the whispering gallery of St. Paul's. I have a fish-pond into which when water is trickling at the upper end, I hear the sound at the lower, 200 feet distant, as if it were close at hand. I remember also, when at school, we used to suspend a large fire-shovel by a string passing over the lower teeth, and amuse ourselvss with the peculiar deep sound which it gave the holder when struck; the ears being close stopped by the fingers.

z. See Climate of London, ii. 327, on the subject. It is probable the experiments detailed in this First Lecture will have required, even with the assistance of a person to work the machine, the full time that it is convenient to allot to the Lecture. Should there have been found a void, it will be easy on another occasion to add a few more from the common books on Pneumatics: but the probability is, rather, that more have been introduced than can be got through in two hours of time.

NOTES: LECTURE SECOND.

a. The loss of heat in a Solar Eclipse of ten digits amounted to eight degrees of the Thermometer : See Climate of London iii. 31. and ii. 314.

b. For a short account of the greatest storm of wind ever witnessed in this country, the date 1703, see Clim. Lond. i. 255, and the references in p. 254. One of rival force, but of less continuance, has occurred since these Lectures were put to press, affecting the West and South of England: the reports of damage done by it have occupied the public prints repeatedly.

c. See instances of the burning of mills by "running amain," in Clim. Lond. iii. 35, 133. One or two cases were mentioned in the papers as occurring in the late tempest.

d. The Southern hemisphere *having its summer while we have our winter;* and the Earth taking up no more than one hundred and seventy nine days, to pass from Equinox to Equinox in that part of its orbit, while in the other (in which we get our summer) it takes one hundred and eighty six days, the consequence is that the warmer half of the year is shortened to the inhabitants of the Southern hemisphere ; *and its atmosphere is on the whole of lower temperature.*

e. Thus, the wind corresponding to our North (or *polar*) wind being with them the South, the *Trade* (our N E.) wind becomes with them the South-east ; and the returning wind is (not the S W. as with us but) the North-west. We shall have to speak in a future Lecture of the opposite qualities, in the two hemispheres, of the winds of like denomination.

f. It is not needful to go as far as within the Tropic, to find a cause for the phenomena here described. Every parallel of Latitude to the South may be capable of elevating, and returning in a superior current, the air it receives from the one North of it:—but the effects must necessarily be felt most extensively, where the heat (which is the proximate cause) is found in the greatest abundance. There are also to be taken

into the account in its details *a variety of local winds*, arising
from the deflection of the aerial stream by coasts and mountains, or from the openings presented by extensive vallies, and
straits or friths of sea; of which it would be impracticable
(were they better known) to treat in this Lecture.

From a comparison of our own winds and weather with a
Meteorological Register kept in *Iceland* in 1810, it appears
that the two islands of Britain and Iceland are usually in
opposite currents—the one being subject to the Polar, while
the other is under the Equatorial current,—and *vice versa*:
See Clim. Lond. ii. 124.

g. The marks on a common globe may be had recourse to,
for the purpose of describing the Monsoons in connexion with
the months they blow in.

h. In the Climate of London, Intr. p. vi. will be found some
practical observations on vanes, and a description of one on the
best construction; with an engraving, which the Lecturer
may copy and exhibit with advantage, on a large Scale.
It goes without oil, and will never stop by freezing.

i. See Clim. Lond. i. 74. with the Tables D, at the end.

k. For great part of a Century the Thermometer in general
use in Britain, but not much used abroad. It is convenient
to have two of these Instruments (now manufactured cheap
and of good quality:) one, chiefly to shew Atmospherical
temperatures, graduated from ten below *zero*, to 120°: another,
for experiments on liquids, &c. from the freezing to the boiling
point of water.

l. The vessels for this experiment should be thin, and so
managed by a previous temporary filling, as that the temperature
of the vessel itself shall not enter into the result: for which
purpose *tin* may be found as convenient as glass.

m. Water has been found to indicate two hundred and
fourteen degrees on one hand, under the highest *natural pressure*, and two hundred and ten on the other, under the lowest;
both in places near the level of the sea. Carried to the top of
mountains, it boils at temperatures lower in proportion to the
elevation of the summit in the vacuum of an air-pump it
boils freely, at a heat easily borne by the hand. Ice, forming or melting, has not been found to change its temperature
from variation of pressure: but water, in order to congeal at
thirty two degrees, requires motion or the access of air; and

when kept very still may be cooled down many degrees below the freezing point; remaining fluid, but liable to congeal in part on the slightest agitation.

n. Several instances occur in books of travels:—but it is matter of too common experience to need that kind of authority for proof.

o. The Mean or average heats of the years from 1789 to 1831 may be found in Clim. Lond. i. 7, 40; and the Extremes of each month, in each year, in the Tables B. B2, at the end of the volume.

p. These calculations of the Mean Temperatures of the different Latitudes will require much correction, especially for the colder climates, when the necessary observations (for there is no end of *theories*) shall have been had and compared. The Mean heat of the latitude of London, by these, is 51·3° —it should be 48·8°: but the estimated Temperatures within the Polar circle must be greatly in error *on the side of warmth.* Capt. Parry's Register, kept from Sept. 1819 to August 1820, between the parallels of seventy four and seventy five degrees N. Latitude, makes the Mean heat of that year *a degree and a third below the zero of* Fahrenheit, or near thirty degrees lower than the Temperature here assigned. The *maximum* of this Register is 60°, the *minimum*,—50°; and the *range*, consequently not greatly differing from our own. But it may have been among the coldest seasons occurring in that part of the globe. See Journal, &c. of the Hecla and Griper: 4to. 1821.

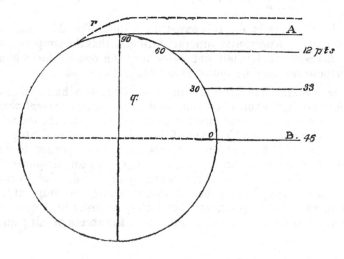

q. The diagram for this purpose, which is represented in figure *q*, may be drawn in strong lines on a sufficiently large scale, for exhibition to an audience.

r. A refracted ray is represented in the figure by the dotted line *r :* it is intended to shew how a portion of the sun's light and heat, which would have passed the Earth in a tangent, is brought down upon the pole by this property of the medium through which it is sent.

s. The Sun was absent from *Winter harbour* (in Captain Parry's first expedition, 1819—20) eighty four days : the twilight, arising from the cause here mentioned, was sensible in the South the whole of the time—they could read the smallest print, by the natural light of the heavens, on the twenty first of December, by *turning the book directly to the South;* and the observed refraction, on the return of the Sun on the third of February, proved that the luminary had been visible to them, in this way, *twelve days longer* in the whole than he would have been by the direct rays alone.

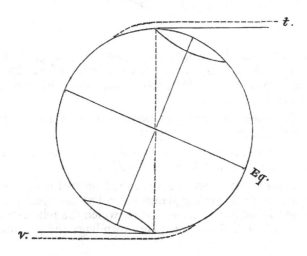

t. u. v. This part should be demonstrated with a pointing rod on a twelve inch globe, duly adjusted :—or, in defect of this, by help of a diagram made from figure *t. v.* The globe will also be helpful in treating of the passage of the Sun to either Tropic alternately, and the effects on the seasons in that part of the globe : of which more in Lecture fourth.

w. It appears then, that but little dependence can be placed on any estimate of a Mean temperature, from observations in which the elevation of the place above the sea is not known.

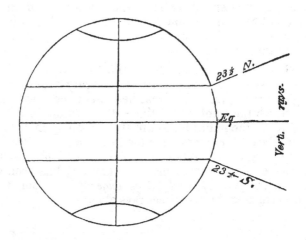

The annexed figure may be copied on a large scale, to accompany the others. It shews the position of the Sun, at the Equinoxes and Solstices, (see p. 32,) and will be referred to hereafter.

x. Of which, instances are mentioned in books of Travels published within a few years past.

y. See the experiments of the industrious and persevering *Wells*, in his work on Dew, published 1814.

z. *Trees* become perfect here at the height of a few inches: there are hardy flowering plants also ; but the chief product of the soil is lichens, such as that on which the rein-deer subsist, and which is found too on our own bogs and mountains.

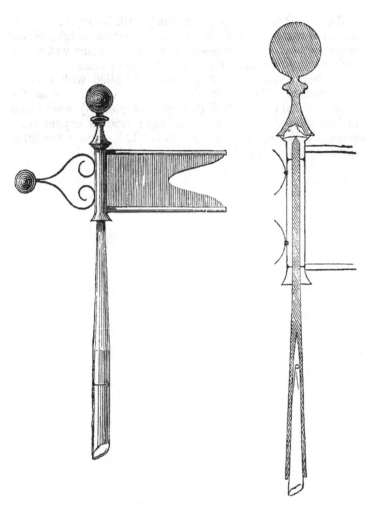

The figure above represents the *Vane* of which mention is
made, p. vii. The following description (it is presumed) will
enable a good workman to construct one from it. A *spindle*
of hard iron, tapering to a point like that of a pencil, (not
sharper,) is mounted on a fir pole of the requisite length,
and of a proportionate thickness. The spindle is received
into a *pipe* of tin or copper, with guides in it: the pipe is
capped with a *ball* filled with lead (to keep it down) having

in the neck below it a piece of *flint*, with its flat or concave surface downward. On this rests the whole weight of the moveable part of the vane—so that the *iron* alone wears, and would not require to be sharpened for many years.

The ball opposite the flag should be filled with lead, to counterpoise the flag: and whatever form is preferred for the latter, it should be of full (not open or pierced) work; and the ball should be mounted on light iron or copper stays presenting an edge only to the wind. This secures the turning of the vane with the least possible motion of the air.

NOTES: LECTURE THIRD.

a. The Climate of that part of England, in which the Metropolis is situate, is not only the one in which the phenomena have been most fully noted, but *the only one in which they have been reduced to a systematic form.* The references to the work in which this has been done will therefore be, of necessity, pretty numerous in these Notes. They will apply altogether to the *Second Edition* of the " Climate of London" in three volumes, 8vo. with plates, 1833: sold by Harvey and Darton, and the other London booksellers.

b. The Mean Temperatures of the Years from 1789 to 1831 are treated of and exhibited on a scale, in the above-mentioned work. Vol. 1, p. 5—10, 40—44.

c. The Lecturer must prepare a diagram on a large scale for the present purpose, from the figure here given: [see the fig. over leaf:] the years from 1789 to 1799, not being insisted on in the text, need not be copied out; they are intended to exhibit a series of Annual Mean temperatures, descending in the scale of heat.

d. Those who incline to study the subject, will find enough to do in reducing the great plenty of Observations, published for various other places, to the form here given. The diagram, or rather map, now to be described, is contained in two plates fronting the Title, in Clim. Lond. There is another form of it, sold also by Darton and Co. on a separate sheet, with letter-press annexed : this may be exhibited to a small company, (with some variation in the terms used in describing it)—but it is not on a scale for large audiences. It is entitled, " *A Companion to the Thermometer for the Climate of London.*"

e. Instances of sudden death, from the effects of a hot Sun on the head, have been reported in the papers in some of our own hot seasons: in more Southern latitudes they are frequent : and in such climates a very light dress is essential to health.

f. The sands in the African deserts have been remarked by Travellers to acquire at times so high a temperature, that the natives, (inured as they are to heat) will not trust their feet on them without *sandals.*

c. Diagram of the Mean Temperatures of the years from 1789 to 1816; as found by observations made in London and its vicinity: Clim. Lond. vol. i. pa. 5. The dotted line continues the London mean.

g. Clim. Lond. i. 31, 35—37. There is reason to apprehend that the quantities of heat thus imbibed by the soil, and transferred from the earth to the atmosphere, differ greatly in amount in different seasons.

h. This instrument, with the manner of using it, is described at p. 55, further on.

i. Clim. Lond. ii. 50.

k. Clim. Lond. i. 19.

l. Clim. Lond. i. 23.

Extreme heat, July, 1808.

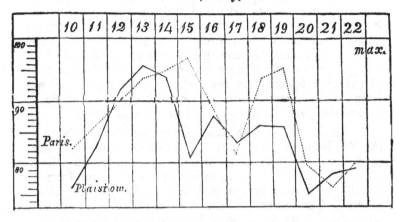

Extreme cold, February, 1816.

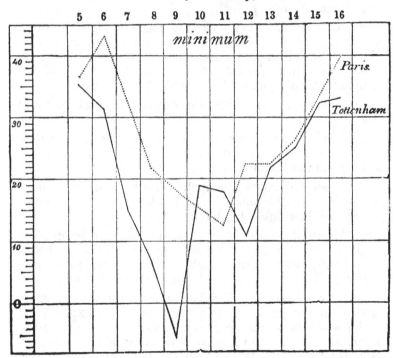

m. In the late expeditions to the Polar sea, the greatest inconvenience was at first found in the accumulation of hoar frost within the ships, (when laid up for the winter), from the action of the intense cold without. A method was however found of collecting the rime produced from the breath of the men, the steam of boiling water, &c. in inverted iron tanks placed over openings in the deck. From these the ice was swept out at intervals, in large quantities; and a sufficient degree of dryness obtained in the internal atmosphere.

n. In the late very abundant snows, which have fallen on the Southern parts (chiefly) of our Island, (and which have produced so much of accident and hindrance to travelling), we have probably been favoured with a discharge from an equally large debt on the cold side of the account, in our Climatic Temperature. The Northern blast, without such a reduction of its power, must have been killing indeed!

o. Clim. Lond. i. 28.   *p.* Idem. p. 35—37.

q. Idem. Introd. p. i. ii.

r. Idem. p. 258—269. See the diagram opposite; from which a large one may be prepared: see also the diagram p. xix, shewing the differences between the London temperature and that of the country, for each month of the year. The Mean temp. of *December*, omitted by oversight p. 53, is 38.71.

s. Idem. p. 29—31. See the diagram forward, p. xviii: and refer to p. 49.

t. The Lecturer should exhibit, for the better explanation of its structure, a figure in outline, on a large scale, of the Instrument itself with the scale affixed.

u. These deductions are applicable also to the variations of the Barometer—some general results of which, thus obtained, will be exhibited in the next Lecture.

v. Ice itself, again, is of so much less specific gravity than the water out of which it is formed, that we see it float, in a thaw, with a considerable portion raised above the surface. This levity may be shown, if ice be at hand.

w. The subject may here be illustrated (if the season or the vicinity of an ice-house permit) by actual experiment, with ice and warm water taken by weight, at a certain temperature of the water; to prove that the heat of the latter is absorbed by melting the ice, and does not shew itself in the mean result.

τ. Mean temperatures of the several months, compared with the sun's progress in declination through the year.

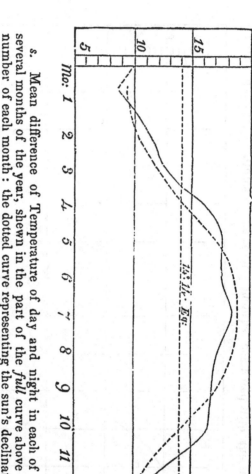

s. Mean difference of Temperature of day and night in each of the several months of the year, shewn in the part of the *full* curve above the number of each month : the dotted curve representing the sun's declination at half-scale. Clim. Lond. i. 30.

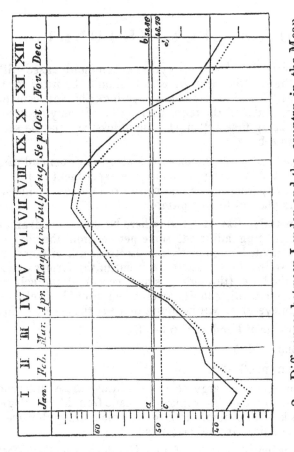

r. 2.   Differences between London and the country in the Mean Temp. of the several months: Clim. Lond. i. 15.   The dotted line represents the temp. in the country.

x. Without which condition it is evident we could have no ice in store for the summer's use.

y. Fish live by respiring water, from which their gills extract the air it contains.   A large pond, iced, may furnish the requisite supply of air to serve them through the frost; a small one, not: in the latter case the fish must die.

z. A large drawing or two, from the Plates given in Parry's Voyages, would much embellish the concluding part of this Lecture.

NOTES: LECTURE FOURTH.

a. The small exhausting syringe, commonly used with a flask for the electric light *in vacuo*, should be fitted for this purpose with a cap to screw on the syringe; having, cemented to it, a stout tube of the requisite length. It may be used in the hand, being first well oiled—the quicksilver may be set on the floor in a basin.

b. The same syringe, connected with the upper end of the open tube and bottle, employed in the experiment *v*, in Lect. 1: but the experiment is much more striking and demonstrative on the plate of the air-pump.

c. A portable Barometer may here be shewn to advantage. The vernier being adjusted, some person from the company should be invited to go with the Lecturer up, or down, two flights of stairs, to witness the descent in the former case, or ascent in the latter (through a small space) of the column: the mercury returning to its level when brought back to the room. An eye-glass will be found to assist the observer, here.

d. Climate of London, i. 66—68.

e. Idem. p. 205.

f. Idem. p. 63—73.

g. Curves on a large scale exhibiting these differences would be instructive: they might be made from Meteorological Registers kept in different latitudes, which are published in various periodical and other works.

h. It is possible that, of two Barometers made perfectly alike or comparable, the one placed within the atmosphere of a city or town might be found always higher, from the weight of the smoke and vapours (at the same hour and *temperature,*) than the one in the clear air of the country. And it might be affirmed (with as great truth as we say, the Earth moves towards the falling body), that the flight of every bird across the field adds to the weight supported, and so to the height of the column.

i. Clim. Lond. i. 298—303.

It may be convenient, here, to take for an example the annexed *figure*, engraved for the above work, of the Barometrical variation through a space of eleven days ; and read the following, referring to the figure *on a large scale* the while.

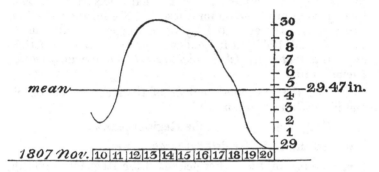

1807. Nov. 10 to 20—part of a windy period of 28 days from New M. to New M., the *mean* of which is 29.47 in. represented by the horizontal line; the *curve* passes through the mean height of the Barometer for each day.

Southerly winds had prevailed from the 28th Oct. with about an inch of rain: the temperature had gone down since the first of the month, from 54° to 44° in the *max.* and from 43° to 33° in the *min.* Thus circumstanced as to the previous weather, a south-east wind on the 10th brought 0.61 in. of rain : after which, the wind changing to N.W., we see the Barom. *rise* an inch (on the mean) in three days, going to about 30 in. A little snow followed the beforementioned rain, and it froze on four successive nights, the first *ice* of the season—the wind N. and N.E. The wind now changing to southward, the Barometer falls rapidly between the 17th and 20th ; *snow* falling on the 19th in considerable quantity, dissolved by *rain* in the course of that day—the whole making about half an inch in the gauge.

We thus have the curve at 29 in.—a depression not only attended with snow and rain, but introductory to a second rain of like amount: after which came a smart frost, with a turbid atmosphere and frequent intermissions of the cold, continued through the greater part of December. The Evaporation during these eleven days was about six-tenths of an inch : it lessened much before the rain, and increased after it. The total rain of the period of 28 days, near London, was 2.83 in.: but at Manchester in the same time there fell about five inches, the mean temperatures being nearly alike. Such is the history of ten days' weather, studied in this interesting manner: a thing which every one possessed of the requisite instruments, and the daily leisure to observe them, may at any time do for himself. See for the rest Clim. Lond. ii. 31. It was found by observations carefully made in 1807 and 1808, near London, *that the rain which fell by day was to that which fell by night as 2 to 3 in amount, nearly.* Idem. i. 110.

k. See the autumnal periods of the Register in vols. ii. and iii. of Clim. Lond. *passim..*

l. Clim. Lon. ii. 7, and the Register *passim.*

m. Idem. in the Register at large, *passim.*

n. Idem. in the earlier periods of the Register, under the head of Evap. in the 2nd volume.

o. Idem. vol. i. p. 312: and 81—88.

p. The annexed is the figure of the scale for an Evaporation-gauge.

See Clim. Lond. *Intr.* xvi. and i. 97.

q. Another form of the gauge, consisting of a cylindrical glass and small five-inch cistern, is here represented. *Note,* that the glass cylinder is graduated also for the *Rain,* but in the reverse direction. For evaporation the full scale is put into the cistern, *and the quantity lost by drying* ascertained by returning the residue into the glass. The *amount of Rain* (through a funnel of five inches) is seen on inspection, when the water is put into the glass. The figure of the rain-bottle and funnel is further on.

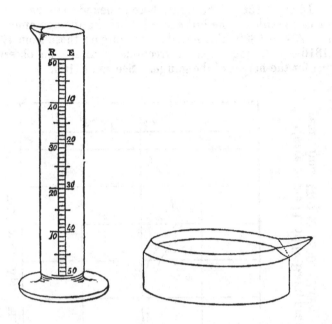

r. Clim. Lond. *Intr.* xiv. and i. 89. See on this subject
the practical observations of Mr. Marshall, in his " Minutes,
&c. on Agriculture in the Southern Counties." 1799. vol. ii.
209, 232. He says in one place, ' I went up to a distant part
of the farm and took the teams into the field, *merely on the
indications of the Hygrometer;* and found the peas [which had
been on the preceding morning *soft*] perfectly fit to be carried.'
He used a hygrometer of his own making ; of stout whipcord,
five feet long, connected with an index traversing a scale on
the segment of a circle. The whalebone hygrometer is a
neater instrument, and will remain longer serviceable.

s. Clim. Lond. i. 99. 108—110.

t. Idem. i. 136. The figure here annexed may be copied on a large scale. The full curve is from the average amounts from 1797 to 1830; the dotted curve, only on those from 1797 to 1816—including some corrections (founded on observation,) for the height of the gauge. See vol. i. 106.

Curve of the Monthly Rain, near London, on an average of thirty-four years, from 1797 to 1830. Average yearly amount 25.13 inches.

u. Clim. Lond. i. 111.   Annexed is the figure of the rain-funnel and bottle—the glass guage is given at page xxiii. back.

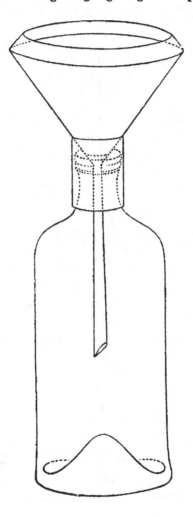

v. Clim. Lond. ' On the Rain,' in vol. i. *passim.*

w. Clim. Lond. ii. 165.

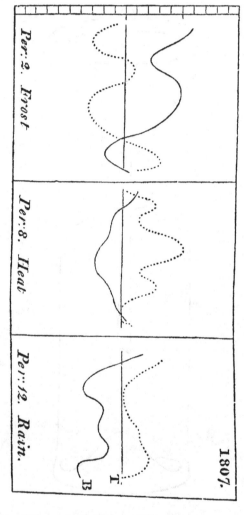

x. Philo. Trans., vol. 123, p. 575—594.

y. Clim. Lond. ' On Periodical Variations,' in vol. i. *passim.*
Philo. Mag. vol. vii. p. 365, &c.

z. Clim. Lond. i. 180—182. 194.   *a a.* Idem. i. 197.

NOTES: LECTURE FIFTH.

a. In ' An Essay on the Modifications of Clouds, and on the principles of their production, suspension and destruction: read before the Askesian Society, in the session of 1802-3; by Luke Howard.' Also the same in *Philo. Magazine,* vol. xvi : ' The Natural History of Clouds ;' in *Nicholson's Journal,* vol. xxx. ; and ART. CLOUD, in Rees's Cyclopædia, both furnished by the same author: with various accounts of the system else-where ; and ' Climate of London,' vol. i. *Intr.* xxxix—lxxii.

b. Clim. Lond. ii. 149 : iii. 225.   *c.* Idem. ii. 21, 22.

d.   Qualis ubi ad terras abrupto sidere *nimbus,*
It mare per medium, miseris heu ! prescia longe
Horrescunt corda agricolis.– VIRG.

e. The Lecturer should be provided here with Indian-ink sketches of the Modifications, taken on an enlarged scale from the plates of the Essay above-mentioned.

f. The red resin called *dragon's blood* is commonly used for the first of these experiments: for the second, there is pro-vided a toy in the optician's shops.

g. Clim. Lond. ii. 7, 17, 49, 146, 278.

h. Idem. ii. 259 : Pub. papers of the season.

i. Idem. ii. 51, 141, 161, 367 : iii. 336, 353, 374.

k. Idem. ii. 14, 24, 85, 113, 69 (there is a good plate of the appearance on the wall produced by this accident, in Journ. de Physique, tome 69), iii. 161, 227, 289, 321.

l. Idem. ii. 113.   This figure gives the appearance of the melted and unmelted portions.

m. Philo. Trans. abr. *passim :* or by the Index in xviii.

n. Idem. vol. ii. p. 309 : iii. p. 18 : ix. p. 653.

o. Clim. Lond. ii. 51, 109 : iii. 289, 291.

p. Idem. ii. 24.

The Lecturer, who would at all times interest and please, should on this subject study the Philo. Trans. and other periodical works, to be found in public libraries; and procure copies on a large scale of the plates; and, where practicable, specimens of the melted iron, shivered wood, &c., to exhibit in describing the accidents to which they belong. The common experiments with the electrical machine, illustrative of the appearances of the spark, star and brush, and of the nature of a discharge with its effects where the conductor is imperfect, should be had recourse to when the weather and situation of the lecture-room favour the attempt. They are not insisted ou in the text, because it too often happens that from various causes this cannot be done to satisfaction.

q. Clim. Lond. ii. 377, 383 : iii. 48. These are described in " Clarke's Travels in Russia," as follows:—" Proceeding towards Celo-petroskia-paulnia, we were surprised by a spectacle similar to that which *Bruce* relates having seen in Africa. We observed at a distance vertical columns of sand, reaching as it appeared from the earth to the clouds, and passing with amazing rapidity across the horizon. Our servant, a Greek native of Constantinople, related a circumstance of a child in the Ukrain, which was taken up by one of these tornadoes, and after being whirled round and round, had every limb broken in its fall." p. 192.

Similar whirlwinds of dust are mentioned as raised in the dry season in India, on the plains of the central and eastern districts of the South Mahratta country. " Not a patch of verdure, not a tree or shrub is to be seen. Clouds of dust are swept along by the parching wind; or huge pillars of it raised up by whirlwinds to the height of a hundred feet, are seen stalking across the plain, or if the atmosphere be calm, fixed for a length of time to one spot." Dr. Christie in Edin. New Philo. Journal.

These winds, as well as the pillars of sand, are mentioned also by *Parke.* " In the afternoon [of March 25, at Benowne], the horizon to the E. was thick and hazy, and the Moors prognosticated a *sand-wind* [which came on accordingly]. It swept along in a thick and constant stream, and the air was at times so dark and full of sand, that it was difficult to discern the neighbouring tents.

"April 7, about 4 p.m., a whirlwind passed through the camp with such violence that it overturned three tents, and blew down one side of my hut. These whirlwinds come from the great desert; and at this season of the year are so common that I have seen five or six of them at a time. They carry up quantities of sand to an amazing height, which resembles at a distance so many moving pillars of smoke." p. 203.

Jackson, in his ' Empire of Morocco', says of the spaces between the *Oases*,—" These stages are very dangerous when the hot and impetuous winds, denominated Shume, convert the desert into a moveable sea, aptly denominated by the Arabs ' El bahar billa maa'—or a sea without water; more dangerous than the perfidious waves of the ocean."

Whole caravans are sometimes thus lost, by the accumulation of ' a mountain of sand;' which sand is also carried sensibly many leagues out to sea." p. 283. 3d Edit.

r. Clim. Lond. ii. 83, 184, 303, 340, 382: iii. 361.

s. The more destructive (probably) of the two *to buildings*, if the descending movement exist.

t. For examples of these Electrical tornadoes, or land-spouts, see Clim. Lond. ii. 136, 257; iii. 70, 123, 253, 283, 295 : also the Philo. Trans. *passim*, and other periodical works.

u. See the Philo. Trans. and Periodical Journals *passim :* and Clim. Lond. i. 220: ii. 181, 273, 335.

v. Polynesian Researches, by William Ellis. 1829. Vol. i. p. 479—486.

w. Account of a Waterspout at sea. In the ' Friend' paper; taken from the Mechanic's Magazine.

x. The Lecturer should by all means be provided with correct Indian-ink sketches of this grand phenomenon (from some of the many plates of it published,) on a scale sufficiently enlarged for his audience.

NOTES : LECTURE SIXTH.

a. The Electrical machine (should the circumstances be favourable,) may be suitably used in illustration here, though it may have been exhibited in a former Lecture : but this seducing toy must not be suffered at any time to supersede the delivery of the text.

b. Philo. Trans. abr. vol. x. p. 525. It is clear from the accounts of eye-witnesses, that the Professor was killed by a stroke of lightning upon the house ; which tore the chamber door off the hinges, split the door-case, stopped the clock, and threw the ashes from the hearth about the room; it is also clear that he received the immediate stroke from one of his insulated iron rods, the index of which he was observing at *about a foot distance from the rod.* This fact should be adverted to at times, as a caution to us, to be careful *how we proceed to question nature* in these her more dangerous operations.

c. Clim. Lond. ii. 13. Such observations as are here described may be made with perfect safety. See the description of the apparatus, Intr. xxv.—and of Read's, in Philo. Trans. vol. lxxxii. pt. 2: and compare the observation *h* under Tab. xviii. (of Clim. Lond.) with the present.

d. e. f. Compare also, here, the facts contained in the ' Essay on Atmospheric Electricity' in Clim. Lond. i. 137—153. The Lecturer may do well to exhibit a copy, on a larger scale, of the plate in these observations ; and explain with its help this part of the subject.

g. Take the following from the Philo. Trans. Abr. as a specimen of the phenomena, confirmed by various other (and some of those very recent) accounts. Mr. Nicholson, teacher of the mathematics in Wakefield, says—" On the first of March, (1774,) about half-past six in the evening, as I was returning from Crofton, a village near Wakefield, I saw in the North-west a storm approaching—the wind which had been strong all the day, setting from the same quarter: and as in the after-

noon of the same day there had been some violent showers of
hail, I made the best of my way to the turnpike at Agbridge.
The air was so much darkened, before the storm began, that it
was with difficulty I found my way [on the bridle-road leading
to the high-road]. When I was about 300 yards from the
turnpike the storm began: when I was agreeably surprised
with observing a flame of light, dancing on each ear of the
horse—and several others, much brighter, on the end of my
stick, which was armed with a ferule of brass, but notched with
using [so as to present separate points]. These appearances
continued till I reached the turnpike, where I took shelter.
Presently after there came up five or six graziers, whom I had
passed on the road. They had all seen the appearance, and
were much astonished: one of them in particular called for a
candle to examine his horse's head, saying *it had been all on
fire, and must certainly be singed!*

"After having continued about twenty minutes, the storm
abated, and the clouds divided, leaving the Northern region
very clear, except that about ten degrees high there was a thick
cloud, which seemed to throw out large and exceedingly beautiful streams of light, resembling an Aurora Borealis, towards
another cloud that was passing over it; and every now and then
there appeared to fall to it such meteors as are called *falling
stars.* These appearances continued till I came to Wakefield,
but no thunder was heard. About nine o'clock a large ball of
fire passed under the zenith, towards the S.E. part of the
horizon. I have been informed that a light was observed on
the weathercock of Wakefield spire, which is about 240 feet
high, all the time the storm continued." Vol. 13, p. 538.

Here are a variety of Electrical phenomena described in
connexion, *as happening together;* and it will be well to bear in
mind this fact as we proceed. See also on this head, Bibl.
Britann. Jan. 1823; and the same in Clim. Lond. iii. 73: and
an article in the *Hereford Times,* of Saturday, Jan. 7, 1837.

h. See Clim. Lond. iii. 74, on the subject of this fact as found
in Pliny, whose text is there given; he makes the corposant a
body of fire, shifting its place like a bird on the ships masts,
and emitting an audible sound. See also in Philo. Trans. abr.
xiii. 35, an. 1770, an account by Capt. J. L. Winn., of a light
and sparks proceeding *for the space of two hours and a half*
from a place, where the electric communication had been inter-
rupted by the accidental breaking of the conducting chain of
his ship, below it. This proves that ships must be frequently

(and even in their ordinary state, or unfurnished with a chain
to the mast-head) conductors to a great extent of the electric
fluid, contained in the clouds and air above, and around them.
And this property may often be the means of saving the vessel
from destruction. But for instances of the latter event see in
the same vol. xii., 157, an account of several ships struck and
damaged, and one wholly burnt; also Clim. Lond., vol. ii. and
iii., and several cases by the index.

The *Philo. Trans.* will also furnish the Lecturer with a
number of curious facts, illustrative of the effects of the electric
fluid (in the case of a stroke on a *building*) among the different
substances it meets with in its course to the earth. He will find
it here making no distinction of sect or party, of property sacred
or profane—putting out the candles, and scattering the wafers,
and spilling the wine, in the Romish church, amid the congre-
gation at Stralsund, vol. i. 526); knocking down the steeple
and dismounting the bells, and breaking and tearing out of
their frames the ten commandments in English episcopal
churches (xi. 113, xii. 126, 610), entering the tabernacle in Tottenham Court Road soon after Mr. Whitfield had built it, (and
on a Sunday too), and doing much damage there and killing a
man. In some other record he may find also an account
of a similar and more dreadful accident at the large Wesleyan
chapel at Stockport, of later time.

But it is in private houses that we hear of the greatest variety
of its exploits. It is capable, it seems, not only of splitting
and boring in all directions, and through all sorts of building
materials, but of sweeping chimneys also, making fire-works
with the soot on the floor, and powder and shot out of the iron
of the bell-wires; of cutting iron, driving nails and drawing
them, shivering glass in one situation and melting it in
another—and, while it breaks six dozen of wine at once under
a vault, 'every botttle as if with a mallet,' leaving the oil in
a bottle on the kitchen shelf untouched, but with the empty
part of the bottle taken off *just above the fluid.*

Nor does it much more respect the gay assembly than the
parties in private rooms: At Naples, Mar. 15, 1773, in the house
of Lord Tylney, and on his lordship's assembly-night, with most
of the Sicilian nobility, and foreign and English ministers of
distinction present, &c. to the number of five hundred, the lightning passed through nine rooms, seven of which were crowded
with parties at cards, or conversing; it was visible in every one
of them, notwithstanding the quantity of candles, and left in

all evident marks of its passage. Most of the company were
electrified, and some of them felt it for several days after—but
no one, it seems, was really hurt save the French ambassador's
servant, who had a black mark on his shoulder, &c., and another who had his hair singed, having fallen asleep with his
head against the wall. It seemed to a Polish Prince to be a
pistol let off near him, and feeling struck, he jumped up,
drew his sword, and put himself in a posture of defence; and
the two philosophers, Sir William Hamilton our ambassador,
and M. de Saussure, happening to be looking different ways
at the time, were so cheated as each of them to make out
the light and explosion to be in front of himself! Vol. xiii. 455.

i. Clim. Lond. ii. 65. The year 1671 seems to have been
memorable in the West of England for ' freezing rains'—which
fell on the 9th of December, and some following days, and so
loaded the trees as to break them down, and make the high-
ways ' unpassable.' I weighed (says the reporter) the sprig of
an ash-tree of just three quarters of a pound, which was brought
to my table; the ice on it weighed sixteen pounds. A very
small bent at the same time was produced, which had an icicle
encompassing it, of five inches round by measure. Yet all this
while, when trees and hedges were laden with ice, there was no
ice to be seen on our rivers, nor so much as on our standing
pools. Philo. Trans. Abr. vii. 37.

j. Clim. Lond. ii. 50, 79, 107, 131, 257, 271, 370; iii. 83,
218, 281, 319. &c. Philo. Trans. Abr. vol. i. 168: iii. 568:
iv. 171, &c.
In 1360, when Edward III. was retreating from Paris with
the English Army, harrassed by want and fatigue, he was over-
taken near Chartres by a most dreadful hurricane, with thunder and lightning, and hailstones so large, as killed instantly
6000 of his horses and 1000 of his best troops."—*English History.* Compare with this Josh. x. 11, Exod. ix. 22—26, and
Job xxxviii. 22, 23.

The French, who suffer much more from hail than we do,
and whose fields are subject to these storms *at particular distances from the mountains,* have a method of blowing up the
cloud with gunpowder; of 'which they explode large quantities
on the hills where it is seen to collect. See Clim. Lond. ii.
14, 142.

k Philo. Trans. Abr. vol. xi. p. 1 and *plate.* Clim. Lond.
ii. 259, 325, 355, and further by the Index.

l. Idem. ii. 291, 351 : iii. 109, 140, 243, and the public prints respecting the dreadful accident at Lewes.

m. Clim. Lond. ii. 389: an observation by Dr. Burney, of Gosport, confirming this.

n. See observations of both kinds of hoar-frost, Clim. Lond. ii. 219: on Rime and its effects, Idem ii. 58, 225 : iii. 104, 136, 347.

o. On the *gossamer,* or webs of the aëronaut spider, see Clim. Lond. i. 252 : ii. 255 ; on the *Corona* and *Halo,* Idem. ii. 102, 126, 265, 281, 319, 390, 396, 401 : iii. 47, 245, and on the manner of the production of both, i. 223.

p. Clim. Lond. i. 222: ii. 331 : iii. 326, 367.   Philo. Trans. Abr. vol. xi. p. 532.

q. Clim. Lond. i. 222: iii. 327.   The diagram mentioned in the text may be constructed from the figure given in the Philo. Trans. vol. 77, p. 46 ; and it will accompany the following description of the appearances, and weather attending them, by Alex. Baxter, Esq., in a letter to Sir Jos. Banks, P.R.S.

"At Fort Gloucester, on the river of Lake Superior, in North America, Jan. 22, 1771.   Last night and to-day the frost has been more severe than at any time this winter.   I was hardly able at mid-day to keep my face to the wind uncovered, though the sun shone very bright, and the sky clear.   In the morning the wind was Easterly, which veered about with the sun to the South and Westward, returning to the East in the evening : a very small breeze.

"A little before two o'clock p. m. observed a very large circle or halo round the sun, within which the sky was thick and dusky, the rest of the atmosphere being clear.   A little more than one-third of the way from the horizon to the zenith was a beautifully enlightened circle, parallel to the horizon : which went quite round till the two ends of it terminated in the halo—where, at the points of intersection they *each formed a luminous appearance* about the bulk of the sun, and so like him when seen through a thick hazy sky, that they might very easily have been taken for him.   Directly opposite the sun was a luminous cross, in the shape of a St. Andrew's cross, cutting at the point of intersection [of its two limbs] the horizontal circle, where was formed *another mock-sun,* like the two mentioned above.

In this horizontal circle, directly half way between the sun of the cross and those at the ends of the same circle [or in the halo,] were *other two mock-suns,*—one on each side. So that in this horizontal circle were five mock-suns, at equal distances from one another, and in the same line [or circle] the real sun, all at equal heights from the horizon.

" Beside these there was, very near the zenith, but a little more towards the circle of the real sun [or halo,] *a rainbow of very bright and beautiful colours*—not an entire semicircle— with the middle of the convex side towards the sun [or down-ward,] which lowered as the sun descended."

These appearances continued in all their beauty and lustre till about half-past two. Then, first the cross went off gradually —afterwards the horizontal circle, by portions at a time: then the three mock-suns furthest from the real, the two in the halo continuing longest. The rainbow began to decrease after all these: and, last of all, the sun's circle (or solar halo), but this was observable at three o'clock, or after it. " The weather continued fine, but the next day was a little softer."

It is very probable the figure accompanying this description would be improved (in accuracy) by making the two legs of the " cross" into *segments of circles,* surrounding the upper mock-suns to right and left; and the " rainbow" into a seg-ment of a circle encompassing the mock-sun at top. The figure wants but another mock-sun, at an intersection (or contact) of the halo with the inverted bow, to make it as full and complete as it seems possible for the appearance to be. There were probably light frozen sheets of Cirrostratus aloft; and a frozen mist resting on the earth below, at the time— both of such tenuity (though containing icy spicula of some magnitude) as not to have been noticed by the observer in their proper character of *clouds.*—[ED.]

r. Clim. Lond. ii. 61 : iii. 35.

s. A fine plate of the falls of Niagara, engraved for the sub-scribers by Lewis, from a painting by Vanderlyn, exhibits a *double rainbow* in its proper appearance in the shower, con-stantly descending from a cloud raised by the violent agitation of the waters. The *iris* may be seen commonly, when the sun shines, in the spray of much smaller cascades. On the irides produced by the spray of the waves in a storm, see Philo. Trans, Abr. vi. 55.

t. The substance of various dissertations on the rainbow

appear to be embodied in an article (under the word) in Rees's Cyclopædia—to which the Lecturer may be referred accordingly, if the original be not at hand. In case of treating the subject optically, the Lecture must be divided into two.

u. For various appearances of the rainbow and its connexion with the weather, see Clim. Lond. *passim,* by the index.

v. Philo. Trans. Abr. i. 73, and the fig.

x. Idem. v. 642.

y. Don Ulloa (in his account of the journey made to measure a degree of the meridian under the Line,) speaks of the ' triple circular iris,' as a phenomenon which greatly surprised him when seen in the mountain deserts of the Andes; but which *frequent observations rendered familiar;* it is thus described, as witnessed on Pambamarca.

" At break of day the whole mountain was encompassed with very thick clouds; which the rising of the sun dispersed, so as to leave only some vapours of a tenuity not cognizable by the sight [as to external form.] On the opposite side to that where the sun rose, and about ten toises distant from the place where we were standing, we saw as in a looking glass the image of each of us—the head being as it were the centre of three concentric irises [of red, orange, yellow, and green] : the last or most external colour of one touched the first of the following—and at some distance from them all, arose a fourth arch *entirely white.* These were perpendicular in the horizon ; and as the person moved, the phenomenon moved also in the same disposition and order. But what was most remarkable, though we were six or seven together, every one saw the phenomenon with regard to himself, and not that relating to others. In the beginning (for it continued ' a long time,') the diameter of the inward iris, taken from its last colour, was about 5½ degrees, and that of the white arch, which circumscribed the others, not less than 67. At the beginning the arches seemed to be of an oval or an elliptical figure, like the disk of the sun, [refracted,] and afterward became perfectly circular.—*Voyage to South America,* vol. i. p. 442. Adams's Trans. 4th edit.

This appears to have been in 1736; in 1768 the same phenomenon was observed by Mr. William Cockin, of Lancaster, (Philo. Trans. Abr. xiv. 639,) and in 1780 in the vale of Clwyd by Dr. Haygarth (Manchester Memoirs, iii. 463). Lastly, we have an account of the same thing, as seen from the mast-head

of a ship in the surface of a fog, in Davis' Straits in 1817, by
B. O'Reilly, Esq.   See his ' Greenland,' &c., 4to. p. 202.

In each of these four accounts, the plate shows the same construction of the luminous appearance: viz., coloured irides
immediately surrounding the shadow (projected on the mist),
of the person viewing it ; and an outer white and distant circle,
or ellipse, comprehending the whole. A drawing might be
made from these by theory, on an enlarged scale, to represent
the phenomenon.

It is worth while, now that the book is before us, to quote
here Don Ulloa's description of the differing sensations of travellers, meeting at a station of intermediate height in these
journies: " At Tarigagua [on the way from Guayaquil on the
coast to Quito at 9000 feet elevation], on the 17th of May, at
six in the morning, the thermometer stood at $1014\frac{1}{2}$ Reaumur,
[59° Fahr.]   And having been for some time accustomed to
hot climates, we now sensibly felt the cold.   It is remarkable
that we here often see instances of the effects of two opposite
temperatures, in two persons happening to meet, one of them
coming from Guayaquil, the other from the mountains; the
latter finding the heat so great that he is scarce able to bear
any clothes, while the former wraps himself up in all the garments he can procure.   The one is so delighted with the warmth
of the water of the river, that he even bathes in it : the other
thinks it so cold that he avoids being spattered with it." Vol. i.
p. 199.

Of the station on Pichincha, above Quito, he says, " We
saw the lightning issue from the clouds and heard the thunders
roll, far beneath us : and whilst the lower parts were involved
in tempests of thunder and rain, *we* enjoyed a delightful
serenity: the winds were abated, the sky clear, and the en-
livening rays of the sun moderated the severity of the cold.
But circumstances were very different *when the clouds rose.*
Their thickness rendered respiration difficult: the snow and
hail fell continually, and the wind returned with all its violence,
so that it was impossible to overcome the fears of being together
with our hut, blown down the precipice on which it was built
or of being buried under it by the daily accumulations of ice
and snow." i. 217.

NOTES : LECTURE SEVENTH.

a. *Baldwin's Aeropaidia*, &c. *8vo.* 1786: Lowndes, Fleet-Street; Poole, Chester. This aëronaut ascended from Chester at 40 min. past 1, Sept. 8, 1785, was 2 ho. 13 min. in the air, and traversed about thirty miles.

He rose with the Barom. at 29.80 in. Therm. 65o—and found the Balloon stationary at Bar. 23.25 in. The air at this elevation was as warm as at the ground, viz. 65o—but he had passed through air *ten degrees cooler* at an intermediate elevation. The modification of cloud which prevailed was evidently the *Cumulostratus*,—the appearances of which from above were extremely magnificent : but the effect here described *of the miniature landscape* was due to a veil of vapours pervading the whole sky, but not visible every where *as a cloud.* On this floor of white he saw at intervals where it was thicker the shadow of his balloon moving, with a brilliant iris at some distance around it.—The space of air, between the denser clouds beneath him and the earth, was lost to his perception, and they seemed to touch the ground. It is worth while (as the book is, I believe, very scarce,) to subjoin his account of the ' world of clouds,' as seen before he commenced his final descent—for he had touched ground, and risen again by throwing out ballast. He thus describes what he saw :—

" The balloon was apparently raised some miles above the surface of a concave shallow plate, or shell—or rather an immense plain, which was in general smooth and well defined. But the dense tonitruous masses, rising here and there above the rest, greatly resembled steep and rugged mountains seen in perspective at different distances, from five and ten to at least an hundred miles. An unvaried deep cœrulean and pellucid azure, without a cloud above, enclosed the novel earth; whose surface, whether valley, plain, or mountain in appearance, seemed as if covered to a prodigious depth by successive falls of snow—driven and polished by the winds, and dazzling to the sight; the sun still shining above all, with white unremitting and invigorating rays. A thunder-cloud of most grotesque form, of superior magnitude, density and brightness—a celestial colouring, whose very shade was a semi-transparent blue and violet purple—remained for several minutes under him."

He descended on Rixton moss, twenty-five miles from Chester, a few minutes before four in the afternoon.

b. This phrase is borrowed from our own version of the Bible, in Jude, ver. 13. It is very expressive of the appearance by night of mere space—into which the ancients no doubt considered the Comets as departing *never to return.*

c. The *lead* colour may be thought to arise from the presence of electric matter in the air; a substance may be observed in thundery weather, separated in beds and patches adjoining the clouds, and even as an atmosphere about them, which does not appear at all like the watery aggregates themselves; but rather as a dry vapour. Is this the matter that becomes inflamed in the thunderbolt (as also in shooting-stars and the larger meteors) waiting in a diffused state its precipitation by the electrical discharge?

A figure of the *Cyanometer* on a scale sufficiently large should be constructed from the one here given. See Clim. Lond. *Intr.* **xxiii.**

CYANOMETER.

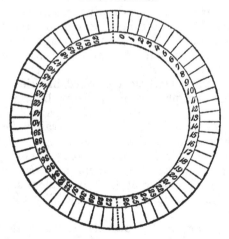

d. Philo. Trans. abridged *by the Index,* in vol. **xviii**: Clim. Lond. by Index ; both under ' Meteor:' and the *Philo. Journals* at large.

e. Any person may convince himself of this, by viewing incandescent bodies at a distance, in the night—and considering how very large a mass of flame would be required to illuminate the whole surface of a country, in the way that meteors do.

The history of Meteorolites has been revived and, minutely treated by Chladri, whose facts are detailed in Tilloch's Magazine : the ancients were well acquainted with the fall of such bodies, *as historical facts :* and Anaxagoras is said to have predicted it in one instance. Again, the masses so seen to fall from heaven were held sacred, and received Divine honours; as in the case of the Ephesian Diana, called in our version of Scripture, Acts, ch. xix., the image which fell down from Jupiter —but there is no word for image in the Greek of the text. See Clim. Lond. ii. 183, iii. 61—and by the Index; in particular (ii. 33.) a minute account of several which fell in the vicinity of Greenfield, Massachusets, Dec. 14, 1807 ; to which a considerable body of testimony of eye-witnesses appears. A very good Epitome of the subject, with the analyses of several of these bodies by Howard, (an eminent chemist of the name, not the author of this work,) may be found in Rees's Cyclop. article *Falling Stones.* They are found to contain the earths *silex, magnesia* and *lime,* with the metals iron and nickel; and where the mass of the fallen body proved to be iron, *nickel* was always found to accompany it.

f. Take for example the following from the Philo. Trans. abr. vol. 7, p. 374. " It being the opinion of divers skilful naturalists, particularly Mr. Fras. Willoughby and Mr. Ray, that the *ignes fatui* are only the shining of a great number of the male glow-worms in England, or of the pyraustæ in Italy, flying together, Mr. Derham consulted his friend Sir Thomas Derham about the phenomenon, being informed that those ignes fatui are common in all the Italian parts." Mr. Derham had himself seen one in England, in a valley between rocky hills, on boggy ground, in a calm dark night. " With gentle approaches he got up within two or three yards of it, and viewed it with all possible care. He found it *frisking about a dead thistle*—till a small motion of the air made it skip to another place, and thence to another and another :" yet he thought it a *fired vapour !*

The account from Sir Thomas states, " That these lights are pretty common in all the territory of Bologna. They are most frequent in watery and morassy ground; and there are some such places where one may be almost sure of seeing them every night, if it be dark: some of them giving as much light as a torch, and some no larger than the flame of a common candle. All of them have the same property in resembling, both in colour and light, a flame strong enough to reflect a lustre on neighbouring objects all around. They are continually in motion, but this motion is various and uncertain. Sometimes they rise up—at others they sink: sometimes they disappear of a sudden, and appear again in an instant in some other place. Commonly they keep hovering about six feet from the ground. As they differ in size, so also in figure; spreading sometimes pretty wide, and then again contracting themselves. Sometimes breaking to all appearance into two, soon after uniting again in one body: sometimes floating like waves, and letting drop some parts like sparks out of a fire."

So far the evidence is conclusive in favour of an insect origin —but now comes the puzzling feature of the case. " And in the very middle of winter, when the weather is very cold and the ground covered with snow, they are observed more frequently than in the hottest summer. Nor does either rain or snow in any wise prevent or hinder their appearance; on the contrary they are more frequently observed and cast a stronger light, in rainy weather."

Thus we are left at last *in the dark*, as we began, with our observations on this singular spectre: and know not whether to call it animal, mineral, aerial or electrical: it is an ignis fatuus, which has left us in the bog of uncertainty and escaped! Compare the notes in Clim. Lond. ii. 121, (where several appearances of it are recorded,) with the attendant circumstances. These (which the author did not himself witness,) appeared in winter, in the midst of an extremely wet season.

g. To do justice to the ancients we may as well here add, that the most absurd interpretations were made of these and other natural appearances in the heavens, in the middle ages of darkness, succeeding to the general reception of Christianity. I have a collection of them (from old histories) mixed with a chronological account of famines, pestilences, wars, &c. (reaching quite down to the Reformation, on the present subject), in a book entitled, ' A general Chronological history of the air, weather, seasons, meteors, &c.,' 2 vols., 8vo. Longman, 1749. From this curious medley the reader may take, as specimens, the following:

" A. M. 3516, appeared a fearful meteor, the whole heavens seemed in a flame: [a grievous plague after it.] 3774. In Tuscany the heavens appeared all in a flame of fire [classed

with ' many dreadful prodigies' of the time.   3791. In Etruria
the heavens seemed to open with a great chasm; [this *chasm* is
probably the black space under the arch, from which the
streamers rise—the light part above being confounded with the
sky: it is here put among ' new and fresh prodigies.'] 3863.
In the sky from E. to W. were seen *armies fighting*. [This
from Pliny, who probably gives it as reported by the Roman
rustics.]

" A. D. 540. Fiery battles, with abundance of blood, were seen
in the air, followed by great inundations, &c.   568. Before
this year that 200,000 Lombardians invaded Italy to dispossess
the emperor of the east, to expel Christianity and introduce
paganism, were seen many prodigious fiery battles fought in the
air; *blood* [to wit the coruscations,] springing out of the earth
and walls.   Again, 570, [before great rains and destructive in-
undations], fiery battles, with blood, were seen over Italy.

788-793. Strange fiery meteors in the air in England, followed by severe famine, and a Danish invasion.   Terrible prodigies in Northumberland—fiery dragons flying—great blasts,
or streamers; soon after followed a severe famine.

867. A cloud was seen hanging over England; one half of
it like *blood*, the other like *fire*.   Soon after the Danes arrived,
burnt, plundered, and murdered without mercy, and carried
multitudes into miserable captivity.

929. A bitterly cold winter.   930. On the 5th of the calends
of March, terrible armies and battles were seen in the air all
night; the noise of the armies, and cries of the wounded, were
distinctly heard !

1098. On the 5th of the calends of October, the heavens appeared all night in a flame.   Another oppressive year *for end-
less taxes and gelds;* [very natural this, *along with the wars !*]
and great rains, which scarce ever ceased, &c.   The like again,
1099.   1106. [Amongst other prodigies and inflictions of
heaven], fiery battles, companies of horse, cohorts of foot,
cities, swords, bloody arms, were seen in the air.

1117.   On the 3rd of the Calends of Jan. and on the 3rd of
the Ides of Dec. the heavens appeared red, and all in a flame
of fire.   Scarcity of corn from the great hail and tempests, and
incessant rains, which ceased little all the year.

1119. [Great rains and inundations.]   Before these rains
the heavens were all in a flame, three hours together after sunset.   On the Calends of Jan. at one o'clock of the night, battles
were fought in the air—first from N. to E. then scattered all
over the sky.

1192—3. [A *rational* notice of the phenomena (from some philosopher) as appearing several times ; and denominated a *meteor.*]

1395. In many parts of England appeared *a thing in the likeness of fire ;* sometimes in one shape, sometimes in another, but in sundry places and at the same time, every night all November and December—but chiefly in Leicester and Northampton-shires. When any body was alone, it *would* go with them : if several were together [to compare notes, and help out each other's judgment] it would go at a distance. To some it was like a wheel turning, all in a flame [as seen overhead] : to others it was like a barrel, throwing out flames of fire at the top [the whole appearance of the dark arch and streamers seen in the north]. To others like a long burning lance, &c. : appearing to all in different forms, it stood when *they* stood, and went when *they* went !

1568. In clear nights were seen in several places of Germany, *two armies in battalia,* brandishing their glittering pikes as if they were ready for a charge : *soon after began the religious war ! Strada.*" So much for 'prodigies.' See also Philo. Trans. vol. 74, p. 227—230, and the *notes* under the passage.

h. The subject is noticed several times incidentally in the publications of Captains Ross, Parry and Franklin ; and in the ' First Journey' of the latter officer to the Polar sea, will be found three articles of considerable length, containing observations by Capt. Franklin, from p. 549 to 569—by Lieut. Hood, p. 580 to 595 ; and by Dr. Richardson, from p. 596 to 628.

The *Experiments,* referred to in the preceding page of the text, are the ordinary ones on the Electric light in a vacuum.

i. This ' black cloud' is clearly the *chasma* of the antients— a space under the great arch at the base of the beams, through which the remote sky appears dark by the contrast. See, for a strong representation of it, the Philo. Trans. for 1836, pt. 1. *Memoranda made during the appearance of the Aurora Borealis,* on the 18th Nov. 1835 : by Charles C. Christie Esq. M.A. with four sketches—p. 31—4, plate ii. and iii.

j. The Philo. Trans. abr., by the Index in vol. xviii. will afford many notices. See also Clim. Lond. vols. ii. and iii. by Index—and the Philo. Journals at large.

k. The fact of its appearance in high southern latitudes has been fully ascertained by our own navigators : but the writer has not his authorities at hand *l.* See on this subject the whole Heathen theogony, with the sacred history and prophecies of the Jews, at large; the apocryphal book of ' Wisdom,' ch. xiii. xiv. xv.; and ' Bryant's Mythology', in 3 vols. 4to. 2nd Edit. 1775.

It may be here remarked that the Lecturer may, by a careful perusal of the whole of the ' Climate of London,' gather more from the work for himself than he will be led to by these references. He may likewise find in this way curves, and tables from which to construct them on a large scale, applicable to the purposes of illustration—with which it was not thought expedient to burthen this little manual.

APPENDIX TO NOTES.

Containing some particulars of the Barometrical variation omitted in note (f), on Lect. IV. and which may be shown on occasion by means of diagrams on a large scale.

In the figure opposite the *perpendicular lines*, placed against a scale of inches and tenths, show the *actual greatest range* of the barometer for each month, in the years from 1807 to 1816, in the neighbourhood of London; being the difference between the highest and lowest observations made in those ten years. The *dotted curve*, which passes through them, shows the *mean range,* or difference between the average of the maxima, and that of the minima, in each month for ten years   The figures in page 65 apply to the full range as seen in the perpendiculars: but by an oversight they are there said, in the italic head, to represent the *average range.*

The dotted curve psesents a very beautiful gradation through the summer half-year; the effects of winds from the north and south being weaker, as the sun gains power by increase of declination; so that the great depressions and great elevations of the column go off together, as the heat increases; and the variation makes an approach towards that of the tropical regions. But in proportion as the cold of winter returns, the effects not of northerly winds alone, but also of the southerly, are more felt; and November, December, and January, present the largest range—which then decreases to the month of March; the mean range of which very nearly returns in October.

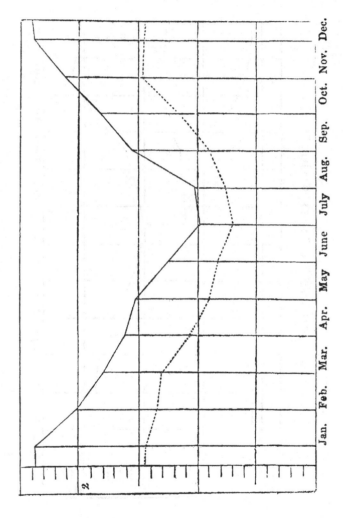

GREATEST AND MEAN MONTHLY RANGE OF THE BAROMETER : 1807—16.

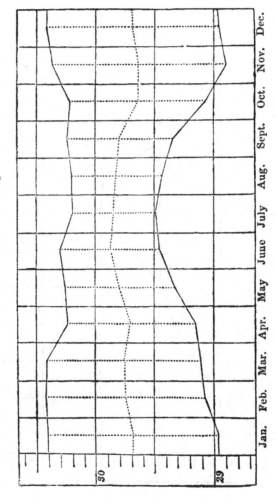

BAROMETER 1807—16 : SEE p. xlviii.

In the figure opposite are represented, *against a portion of the barometrical scale*, the following particulars, deduced from the ten years observations from 1807 to 1816.

1. By the full curve above passing through them, the average of the ten maxima, or highest observation, in each month in those years.

2. By the full curve below, in like manner, the averages of the ten minima, or lowest observations.

3. By the dotted perpendiculars, the *mean range* in its place in the scale, which was exhibited after a different manner in the former figure.

4. By the dotted curve passing through the middle point of each of these, the actual additions to the weight of the air in the summer months, from vapour sustained by the heat; and independent of the effects of opposite winds above-mentioned. See page 64.

END OF NOTES.

ERRATA ET CORRIGENDA.

PA.	LINE.	
15	19	Ocean *read* seas.
41	14	Virgo—Scorpio.
53	4	from bottom . Supply Mean Temp. 38.71°.
65	11	Average—Actual.
110	16	Spicula—spiculum.

J. LUCAS, PRINTER, MARKET-PLACE, PONTEFRACT.

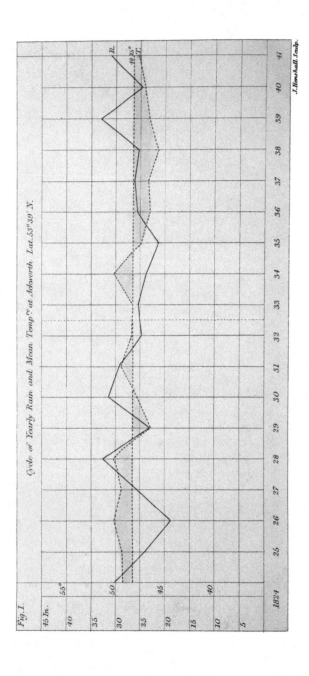

Fig. 1.

Cycle of Yearly Rain and Mean Temp.ͬ at Adsworth. Lat. 53° 39′ N.

J. Henshall. Sculp.

A

CYCLE OF EIGHTEEN YEARS

IN

THE SEASONS OF BRITAIN;

DEDUCED FROM

METEOROLOGICAL OBSERVATIONS

MADE AT

ACKWORTH, IN THE WEST RIDING OF YORKSHIRE,

FROM 1824 TO 1841;

COMPARED WITH OTHERS BEFORE MADE FOR A LIKE PERIOD
(ENDING WITH 1823) IN THE VICINITY OF LONDON.

BY LUKE HOWARD, Esq., F.R.S., &c.

WITH FIVE PLATES.

LONDON:

JAMES RIDGWAY, PICCADILLY;
HARVEY AND DARTON, GRACECHURCH STREET;
BAINES, LEEDS; LUCAS, PONTEFRACT.

1842.

A CYCLE OF EIGHTEEN YEARS IN THE

SEASONS OF BRITAIN, &c.

THE fact of a periodical revolution, bringing alternate warmth
and coldness through successive trains of seasons in our variable
climate, is now ascertained beyond controversy; and it becomes
in consequence an important object, to ascertain the nature and
extent of these changes, and their effects on our *agriculture* especially; that we may the better avail ourselves of the favourable,
and provide against the adverse.

In my account of the climate of London, first printed in 1818
—1820, I gave a view of these changes on the basis (which my
observations then seemed to present) of *alternate periods of seven
and ten years*, the former *ascending*, the latter *descending* in the
scale of heat. I then admitted, from appearances, the probability of spaces between these successive periods not agreeing
with the rule abovementioned, and answering to the "intercala-
tions" of an imperfect calendar. Having since pursued the subject further, I find these spaces, or interposed years, to be necessary parts of the scheme at large; which now resolves itself into
a cycle of eighteen years, in which our seasons appear to pass
through their extreme changes in respect of warmth and cold,
of wet and dryness. I have given an account already to the
Royal Society of my views on this subject, as regards the seasons
near London, and what relates to the periodical variations of the
barometer from year to year in this neighbourhood is inserted in
the Transactions*. My present object is, to bring in confirma-
tion of these views *the facts of a new period, observed in a new
locality,* and that differing so considerably in latitude from the
former, as to justify the inference that the periods are not con-

* Philosophical Transactions for 1841, Part. II.

fined to any part of our island, but will be found, variously modified, in all.

Referring to the papers abovementioned for a variety of facts relating to atmospheric periodicity, stated in a more elaborate way, I shall here briefly analyse the results of the Ackworth register, and apply them to my object; saying little about the *barometer* however, because the present observations on this instrument, however constantly made from day to day, have not the comprehensive character of those insisted on in my former papers; *which were taken from the face of a registering clock.* The Tables annexed to this paper, then, comprise the *Results of a daily meteorological register,* kept at my instance, and with instruments furnished by myself, *at the Friends' Public School in Ackworth.* I have observations, not so continuous, made at my own residence there; by collation with which in many parts I have satisfied myself that I can depend on *these*, for the purpose to which they are here applied, of deducing the differences of seasons from previous and subsequent ones of like denomination, *by comparison with each other.*

As on former occasions, I shall have recourse to *diagrams,* which present to the eye with ease what figures would convey but slowly to the mind. The *dotted* curve, or flexuous line in fig. 1. shows the variation from year to year of the mean temperature, or *average heat* of the year, the *Mean of the climate* (or average of all the observations of these eighteen years) being 48°·152. The nine years from 1824 to 1832 average 48°·952; the nine years from 1833 to 1841 give 47°·352. The difference of 1°·6 is about equal to the difference in warmth between Ackworth, N. lat. 53° 38′ 57″, and London. I therefore call the former nine the *warm*, and the latter nine the *cold* years of the cycle. The curve shows palpably the bulk of the years of high temperature on the *right*, and of those of low temperature on the *left* of the dividing line, but with two striking exceptions. There is a very cold year, 1829, among the *warm*, and a very warm year, 1834, among the *cold*; and these considerably reduce the difference between the two averages: the comparison or contrast holds best, therefore, *among the years in detail.*

The *full* flexuous line in fig. 1. shows the variation, from year to year, of the total rain collected by the gauge in each. It is not here as with the temperatures; the amount of *Rain* is balanced, or nearly so, in each nine years. Thus out of 472·93 inches fallen in the whole cycle, 238·60 inches appear to have fallen on the *warm*, and 234·33 inches on the *cold* side, making the annual

averages respectively 26·51 and 26·04 inches, nearly; which is
about *an inch more* on the whole *per annum* than is found to fall
near London—the level being at the ground, in both. If we
now look through the *curve* (I beg pardon of mathematicians for
applying the term to such a line,) we shall probably be first
struck with an extreme of dryness (1826) followed by an extreme
of wetness (1828) on the *warm* side; then, with a gradation
from *very* wet (again following very dry) in 1830, to *very* dry
in 1835; and this again mounting by steps to extreme wet again
in 1839. In fact, ten years, from 1830 to 1839, show a gradual
decrease and again an increase of rain, protracted through the
half-cycle, while eight years from 1840 to 1829 (passing thus
back to make the cycle) show repeated and more extended oscil-
lations, performed in shorter times; yet with results so nearly the
same, that the first set of years, here specified, show an average
rain of 26·36 inches, while the second set average 26·16 inches.
Again, on comparing rain with temperature, we find 1826 in the
extreme at once of *warmth* and *dryness*, and 1839 in those of
wet and *coldness*: but 1828 (in the extreme of wetness) is equal
in *heat* to the dry 1826; and 1829 is both *dry* and very *cold.*
The quantity of rain, therefore, is not regulated by the *temperature* of the year: we may get it with *heat*, brought by winds
highly vaporised from the tropic; or with *cold*, from the condensation effected by the approach of northern air to our own
atmosphere, previously charged with vapour to the full; and
the dryness of 1829, with so much of *cold*, may have been the
result of the great deposition of rain in the previous season.
The only rule, then, that prevails throughout, seems to be *com-
pensation*; a wet year against a dry one, &c., and so of whole
runs of seasons; and we must examine the *winds* for the cause.

This I have done in my paper before the Royal Society alluded
to above; but the subject, complicated as it is with the moon's
intricate orbit, and her varying influence on the currents that
sweep these islands, is much too extended for the compass of
this paper. We have to do, here, not so much with causes as
with the effects they produce; a knowledge of which must necessarily go before the other. I shall proceed therefore at once,
from the review of the rain and temperature of *whole years*, to
an analysis of the distribution of these through the *several months*
of the year; which will let us probably into the secret of the
difference, *under equal quantities of rain*, of the warm from the
cold side of the cycle; as regards the most important of its effects,
the fruitful or unfruitful character of our seasons.

The full flexuous line in fig. 2. presents the monthly rain, in
its total amounts under each month, for the *nine years* 1824 *to*
1832, or warm period; the dotted curve, the same for the *nine
years* 1833 to 1841, or *cold* period of the cycle. I shall compare these with the temperature in another figure; the present
object is to show the remarkably different *distribution* of the
like *quantity* of rain, in each. The warm period, then, shows
December barely dry, its rain being three-tenths of an inch below
the mean betwixt the driest and wettest months, which is at 19·59
inches. *January, February* and *March* are in the extreme of *dryness*. In *April* of this period, the rain suddenly mounts up to a
point full three inches above the mean, and descends again three
and a half inches below it, to make us a fine *May*. The months
of *June* and *July*, though high in the scale of rain, have the advantage (the former considerably) in dryness of those of like denomination in the cold period. But in *August* and *September* we
see the case reversed; the amounts of rain in the warm *exceeding*
(by about five inches in each) those of the cold period. *October*
is wet in both nearly to the same degree; and by the bye I may
here observe, that the rains in this month fall mostly by night;
verifying a remark I heard many years since from a friend, that
"there are always twenty fine *days* in October." *November*,
though wet, is drier on the warm side by a quantity exceeding
two and a half inches on the cycle: of *December* we have treated
as regards this side—to turn now to the other.

In the cold period, *December* is very dry; which dryness is
ominous of wet in *January, February* and *March*; carrying the
rain in these to an excess over the former period of four and a
half, seven and three quarters, and three inches, respectively.
To these succeeds *a dry April* (perhaps the worst feature for
agriculture of the whole), the rain falling short by *nine inches* of
the former total. This want is hardly supplied by the more
copious rains in *June*; for they come usually accompanied by far
too much of *wind*, in the summer of the cold period. *August* and
September (contrary to what the inexperienced would predict)
are much the *drier* months in the cold years: yet the lower temperature and cloudy sky counteract the early ripening of the
crops. *November* mounts up, in this period, to the second place
for wetness (it has the *first* in the neighbourhood of London)
and often completes the destruction of what the drier weather of
the preceding months had come too late to enable us to save.
I shall not need (were I more capable of the office) to point out
to the experienced farmer the advantages and disadvantages *to*

him of these different arrangements. They are the work of the all-wise Creator; ordered no doubt for the best *on the great scale of things.* It is for us, instead of vainly wishing them otherwise, to turn to the best account (which we surely may, with the helps derived from accurate observation and a full record,) the opportunities they present. The results in figures, from which the curves I have hitherto treated were laid down, will be found in Tables in the following pages.

Fig. 3. gives the rain under each month, *for the whole eighteen years,* represented by a full line; in connexion with the average temperature of each month for the like series of years, in a *curve* (which may be so called without a qualifying remark) corresponding as nearly as those in my 'Climate of London' with *the curve of the sun's declination*; which I have placed, in a fine dotted line, in connexion with both. This diagram is instructive, as regards the average increase of our rains as the sun approaches us from the south, and their falling off in quantity as he recedes towards the tropic of the other hemisphere: and it serves well for a test *of the completeness of the account of temperature through the cycle*; the solar and the thermometric curve agreeing in form, with the like exceptions, as about London; of an accelerated rise in the spring and a retarded fall in the autumn; which I have shown in my work above mentioned to be the necessary conditions of our annual variation. The results in figures belonging to this beautiful curve will be found in the Tables further on.

In figures 4, 5, 6 and 7, I have given the rain and temperature *of the four seasons of the year,* through the cycle here treated; the numbers belonging to which will be found at the foot of each yearly table, 1824, *Winter* (Jan. to Mar.), &c. at the end. To have treated these curves in detail would have involved a repetition, in substance, of much of what has been said upon the monthly averages. They will be found upon examination to present many curious gradations and coincidences, in the increase and decrease of the rain and warmth of each season: they show also at a glance the difference, in each, between the warm and cold sides of the cycle: also what summers were particularly warm, and what winters cold; with the nature of the changes, from year to year, throughout, in our most capricious season, *the spring.* As to *rain,* we see at once that the wet of 1824 and 1827 occurred chiefly in the *Autumn*; as did also the drought of 1826 and 1829: that, in the very wet 1828, the rains fell chiefly

in the *summer*, so that we had potatoes of a pound and a half in
weight, and the crop unequalled in the whole cycle: after which
the supply of this season fell off gradually to 1832, was renewed
in 1834 and balanced then by a dry *autumn*. In 1839 we see
summer and autumn wet; in 1840, both seasons dry: but the
most remarkable feature, perhaps, of the whole is the opposition
of the curves of winter and spring from 1832 to 1841 ; the former
plainly robbing the latter as it advances in the fore part of the
period, and restoring the amount in the latter part. It is to
show this peculiarity that I have placed the curves in the order
they are found in, the winter one lowest. The rains of the
summers 1835 to 1838 appear to be given away partly to the
autumn, partly to the winter; but it is needless further to anticipate the reader in his remarks.

The yearly *mean heights of the barometer* for the cycle, taken
at Ackworth School, are as follows:—

1824.	29·75 inches.	1833.	29·81 inches.
1825.	29·93	1834.	29·98
1826.	29·91	1835.	29·86
1827.	29·84	1836.	29·73
1828.	29·83	1837.	29·80
1829.	29·84	1838.	29·76
1830.	29·81	1839.	29·78
1831.	29·82	1840.	29·84
1832.	29·93	1841.	29·70

Mean of warm period 29·851 inches.   Mean of cold period
29·807 inches.

The mean of the whole cycle is 29·829 inches. It appears,
then, that the warm side has the higher average of pressure by
0·44 in. This is consistent with my former observations, as to
the greater mean height belonging to *the warmer portion of the
year*; which the reader will find in my ' Climate of London,'
vol. ii. p. 281, 1st edit. ; vol. i. p. 207, 2nd edit. And the same
reasons, which there apply to single years, will be found here to
suit periods also: for the period from 1815 to 1823 averaged
at *Tottenham* 29·788 inches, and that from 1824 to 1832, 29·830
inches; giving thus an excess of pressure to the warm years
over the preceding cold ones, of 0·42 in. My clock takes in the
greatest depressions, and makes the mean lower than it is obtained in the ordinary way: so that these averages of the School
and Tottenham, though comparable respectively *at each station*,

are not so with each other.   As it is, they agree with remarkable nearness in the amount of excess of the warm period.

The *yearly mean temperatures* for the cycle at Ackworth are as follows :—

1824.	49·19	1833.	48·36
1825.	49·45	1834.	50·08
1826.	50·25	1835.	47·31
1827.	49·19	1836.	46·52
1828.	50·26	1837.	46·66
1829.	46·64	1838.	45·79
1830.	47·58	1839.	46·69
1831.	49·57	1840.	47·02
1832.	48·44	1841.	47·74

Mean of warm period 48°·952.   Mean of cold period 47°·352. Mean of whole cycle 48°·152: excess of first nine years 1°·60.

If we leave out 1829 on one side, and 1834 on the other, and average the remaining years, they come out thus: the first eight years 49°·241, the latter 47°·011: difference 2°·230.   Now the nine years from 1815 to 1823 average in the neighbourhood of London 49°·065, and the following nine years, 1824 to 1832, 50°·137: the difference in favour of the warm period being 1°·072. I might go on to prove another warm period, antecedent, and ending with 1814, averaging 49°·383; but the locality (Plaistow) differs, and the observations cannot so properly be brought into the comparison, though they show a small excess.   Moreover, the method of computing by whole years and calendar months may bring out inaccurate results in an extended examination. I have proceeded, in my paper now with the Royal Society, on the solar year and the lunar divisions, which are more agreeable to nature; the cycle of eighteen years depending on these principles.   The reader will find the subject of periods of temperature largely treated in my published work above mentioned; which, *as regards its details,* is not in any degree superseded by the present paper.

The *rain* of the cycle at Ackworth is as follows :—

1824.	30·51 inches.	1830.	31·58 inches.
1825.	24·22	1831.	28·37
1826.	18·74	1832.	24·94
1827.	25·04	1833.	25·06
1828.	32·35	1834.	23·74
1829.	22·85	1835.	21·19

1836.	25·21 inches.	1839.	33·16 inches.
1837.	25·39	1840.	24·75
1838.	25·02	1841.	30·81

Rain of the warm years 238·60 inches; of the cold years 234·33 inches. Rain of the whole cycle 472·93, or *per annum* 26·27 inches; the warm side averaging 26·51 inches, the cold 26·037 inches. I have found cause, on examining into past periods, to conclude that the small excess of rain here found on the warm side is not a constant result; but that the *cold* may sometimes be the wetter. The main point affecting our harvests appears to be the different distribution of the rain *within the year* in each period, to which we now proceed: but first of the whole cycle.

The following are *the total amounts of rain* for *each month* of the year, through the cycle from 1824 to 1841; to which are annexed the monthly averages of heat, or *the mean temperature of each month* for these eighteen years. The reader will see, on comparing the latter with the curve of the sun's declination (see fig. 3.), that the warmth and coldness, thus averaged on a cycle, follow pretty strictly (though at some distance, as I have shown in the 'Climate of London,') *the proportionate height of the sun* through the seasons; ascending thus to July and descending again to January: the rain itself being somewhat similarly affected, but with a diminution of quantity in the spring months, and an addition in the autumnal, *depending on other causes*. I have treated these also in my work above mentioned, under the head of "Rain," to which I must here refer.

	Rain.	Mean Temp.
January	29·88 inches.	35·734
February	31·16	38·176
March	25·97	41·600
April	36·49	45·852
May	32·99	51·699
June	45·96	57·921
July	54·52	60·717
August	44·81	59·513
September . . .	48·65	54·947
October	43·46	49·466
November . . .	46·96	41·786
December	32·08	39·856
Total	472·93	Mean of year . . 48·105

All which the reader is requested to compare with the curves.

Now, for the distribution to the several months, in the *warm* and in the cold *period*, the following are the results in figures :—

	1824–1832.	1833–1841.
January	12·67 inches.	17·21 inches.
February	11·75	19·41
March	11·51	14·46
April	22·75	13·74
May.	16·11	16·88
June	21·58	24·38
July.	26·85	27·67
August.	24·77	20·04
September . . .	26·98	21·67
October	22·17	21·29
November. . . .	22·17	24·79
December. . . .	19·29	12·79

Total in the warm years } 238·60   234·33 { Total in the cold years.

Of these quantities we have, on the *warm* side, for

January, February, March	35·93 inches.
April, May, June	60·44
July, August, September	78·60
October, November, December	63·63.

And, on the *cold* side, for

January, February, March	51·08 inches.
April, May, June	55·00
July, August, September	69·38
October, November, December	58·87.

The reader will be pleased to compare the numbers with the curves in fig. 2. I have already made such remarks on these as may suffice for his help.

On bringing into comparison the curve of the rain-cycle, 1824 to 1841, with that for the preceding cycle, 1806 to 1823, in the neighbourhood of London, the following differences appear. At *London*, January, in the whole eighteen years, has 10 inches *more* rain; March, 5 inches *more*; April, 7 inches *less*; May, 6 inches *more*; June, 18 inches *less*; and July about 10 inches; August and September have respectively 14 inches and 20 inches *less* at London. The two following months nearly agree in amount with Ackworth (as does February), but December was 20 inches drier at London. With these differences, the two curves still show the common feature of a general accordance with the sun's declination *through the fore-part of the year*: but this is done away in the London curve, afterwards, by the dryness of August and the extreme wetness of November and December. Were the several cycles of amounts of rain, which we now possess, averaged for each year upon nine successive years, and thus reduced to easy curves (as I have done with the barometrical heights), these would undoubtedly present some striking features, leading to the true theory of our rains in Britain; but this is more than I can here undertake to produce.

The following are observations made, with one of my gauges, by my friend Thos. Fowler, Esq., at Tottenham, for seven years past, placed in comparison with those of the school at Ackworth.

		in.		in.
1835. year's rain at Ackworth	21·19		At Tottenham	24·43
1836.		25·21		29·76
1837.		25·39		20·61
1838.		25·02		23·13
1839.		33·16		28·64
1840.		24·75		21·79
1841.		30·81		31·34
Total in seven years......	185·53			179·70

The annual averages thus obtained are for Ackworth 26·50, and for Tottenham 25·67 inches. If we compare the amount, throughout, we shall see the *wet* years coinciding, except in 1836; the *dry* years set against *mean* ones, each way; and the total amounts affording each an average very near to that of the station itself: the two mean years, 1836 and 1837, at Ackworth, balance the wettest and the driest that occur at Tottenham. The whole confirms a remark I have often made in passing to and fro, that we seldom see the Trent and the Thames full with rains, together; their estuaries communicating with different portions of the surrounding sea, and the respective districts that feed their springs requiring somewhat different winds to furnish them with vapour.

After acknowledging my obligations to the present and former Superintendents, and the successive Clerks, of the Institution at Ackworth, it is proper I should caution my reader against *expecting too much* from the information here presented to him. Should he look for the *same mean temperature*, and the *same amount of rain*, in each returning year of the coming cycle, as are found recorded of a corresponding one in the past, he will probably meet with frequent disappointments; and this more especially in a locality somewhat different. We are yet far from being able to predict seasons in meteorology with the like certainty of date as the astronomer does the coming phænomena of the heavens; and it is even possible that, from the very nature of the causes concerned, we may never arrive at this. The judicious observer, finding certain facts fully ascertained and clearly noted for him, *will know how to make use of these for himself*; and by watching their occurrence in detail, making notes as he proceeds, will endeavour to feel his own way towards the future; independently of empirical and fallacious *predictions*. This is the kind of service which I expect my present labour to render to the country; besides gratifying a reasonable curiosity as to the past. We do not expect to become skilful in other arts without a due share of study and practice; but we seem to forget this self-evident truth, when we take up that of foretelling the weather. The facts here detailed cannot fail to be useful to such as will be at the trouble to examine and compare them, though the inferences they may draw from them should differ. And admitting only that, in the course of years here treated, we experience in succession the various degrees of warmth and coldness, of rain and dryness incident to our climate, it must needs help the farmer, the market-gardener, the planter or nurseryman, the grazier, the sheep-master, to have before him such an approximation to the times and order of their occurrence.

A comparison in detail of the warmth and rain of successive seasons with *the yield and quality of the harvest* belonging to them, would have formed, perhaps, a desirable appendage to the *meteorology* of my paper. But I have not before me the materials for this: and to have waited to collect them would have caused me much delay in publishing it, which I have reasons for wishing to avoid. The subject, moreover, viewed in this light, is a very copious one. It is not alone the grower of *corn* that is interested in a warm or a cold, a dry or a wet season; there are a variety of rural occupations, and of trades and manufactures dependent on them, equally concerned in these. And we have the satisfaction of thinking, that these are severally the best judges of their own needs and occasions; that each will be best able to avail himself to purpose of the meteorological facts, established for him by the labours of men of science.

There is a class of persons, however, to whom the paper may be immediately acceptable, and possibly also useful in regulating their future plans. The poor invalid may be soothed, and those of delicate constitutions encouraged, by the immediate prospect of a nine years' run of seasons having, with little exception, *the higher temperature of our climate*. It may be the means of inducing these to make trial at least of one or two of these, before they resort to other skies more favoured by natural position, but extending over countries far less desirable as residences to a truly British mind. And medical gentlemen, when they have read and considered what is here laid before them, may find arguments in it to strengthen such a conclusion.

At the present moment, when we are debating with earnestness the question of our corn-importation, and while some are looking for an absolute scarcity, it cannot be uninteresting to public men to know what, in a scientific point of view, are the bearings of the question of *climate* and *seasons* on this other and more tangible one. That they are on the whole highly favourable, will, I hope, be an announcement giving pleasure to both sides: yet here, also, I must put in a word of caution. Plenty, I have shown, as regards our grain-crops, is dependent partly on a sufficiency of *warmth on the whole year*, partly on *rains administered in due season*. But we *may* have a season or two to go through, in which, though our personal comfort may be increased by dry and hot weather, the crops may not be so well fed, or the sheep so well fatted, as we would have them. But we abound in resources (through the goodness of God to his creatures), and may cheerfully proceed to do our duty and make use of them, in reliance on the Divine Providence; happy also if, through His grace, we have learned to feel for the distress of others, and do for them *in all respects* that which we would they in like circumstances should do for ourselves. This is the great lesson which many of us have yet to learn, more perfectly, to become more safely *proud* (in a good sense of the term) of the name of *Christian*. My countrymen will, I am sure, not be offended by this hint, from one now nearly of an age to be released from all public engagements; but who finding the matter before him, and apprehending a public service in it, could not well forbear the present.

Tottenham, Middlesex,
  March 31, 1842.

Tables of the Barometer, Temperature and Rain, with Notes on the Seasons.

Mean height of the Barometer, Mean Temperature and depth of Rain in each month, at Ackworth, Yorkshire, through a Cycle of Eighteen Years.

	1824.			1825.			1826.		
	in.	in.	in.	in.	in.	in.	in.	in.	in.
January	29·90	39·53	0·60	30·09	39·03	0·80	30·00	32·32	1·05
February	29·74	39·36	0·89	30·04	39·14	0·34	29·78	42·84	2·23
March	29·71	40·39	1·82	30·08	40·80	1·28	29·92	41·82	0·45
April	29·79	44·89	1·22	29·94	47·22	4·56	30·02	49·95	1·67
May	29·91	50·52	1·10	29·92	52·11	2·52	30·03	51·19	0·79
June	29·84	56·28	4·59	29·89	58·17	1·53	30·20	63·96	0·37
July	29·91	62·40	0·43	30·09	62·80	0·76	29·91	64·72	1·65
August	29·86	59·60	1·67	29·84	61·58	3·28	29·89	64·90	1·00
September	29·76	56·60	6·36	29·83	59·81	0·85	29·85	56·90	5·10
October	29·54	48·49	6·07	29·81	52·32	2·48	29·75	51·84	1·31
November	29·45	43·75	2·25	29·87	40·52	2·86	29·78	40·76	1·87
December	29·66	48·54	3·51	29·71	39·92	2·96	29·80	41·76	1·25
Whole year	29·75	49·19	30·51	29·93	49·45	24·22	29·91	50·25	18·74
Winter (Jan. to Mar.)...		39·76	3·31		39·66	2·42		38·99	3·73
Spring (Apr. to June)...		50·56	6·91		52·50	8·61		55·03	2·83
Summer (July to Sept.)		59·20	8·46		61·39	4·89		62·17	7·75
Autumn (Oct. to Dec.)		46·93	11·83		44·25	8·30		44·79	4·43

NOTES. 1824.—The driest *January* and *July*, and the wettest *September* and *October*, of the cycle of eighteen years : compare these months in 1825, 1826, 1827, 1830, 1833, 1835, 1841. The mildest and the wettest *December* : compare the remaining years.

1825.—The wettest *April* of the cycle : compare 1828, 1830, 1833. *September* the driest but one : compare 1832. The driest *February* but one : compare 1832.

1826.—The driest *year* of the cycle, and making with 1828 the two of highest temperature : it has by far the hottest summer : compare July—September in the remainder. The driest *June* in the cycle, and the warmest : compare 1827, 1828, 1837. The *January* among the coldest : compare forward.

Mean height of the Barometer, Mean Temperature and depth of Rain in each month, at Ackworth, Yorkshire, through a Cycle of Eighteen Years.

	1827.			1828.			1829.		
	in.	°	in.	in.	°	in.	in.	°	in.
January	29·78	35·05	1·58	29·87	40·11	3·51	29·82	32·11	0·65
February	30·04	34·53	0·95	29·77	40·97	1·11	30·02	38·52	1·79
March	29·55	42·98	2·85	29·83	44·09	1·58	29·85	38·81	0·32
April	29·92	48·13	0·78	29·72	46·93	4·01	29·47	44·22	2·94
May	29·70	52·78	1·67	29·84	53·38	1·82	30·00	53·76	0·36
June...........	29·85	58·52	0·66	29·93	60·10	1·27	29·95	58·47	2·08
July	30·02	62·71	1·39	29·64	61·44	9·48	29·71	61·45	3·00
August.....................	29·97	58·39	2·79	29·82	59·74	1·28	29·75	57·68	5·19
September	29·95	56·83	1·76	29·92	56·66	2·76	29·68	52·41	3·27
October	29·72	53·48	5·81	30·01	50·00	0·92	29·91	47·32	1·21
November	29·94	43·56	2·56	29·81	45·00	2·77	29·93	40·99	1·27
December	29·69	43·27	2·24	29·82	44·67	1·84	30·00	33·96	0·77
Whole year	29·84	49·19	25·04	29·83	50·26	32·35	29·84	46·64	22·85
Winter (Jan. to Mar.)...		37·52	5·38		41·72	6·20		36·48	2·76
Spring (Apr. to June)...		53·14	3·11		53·47	7·10		52·15	5·38
Summer (July to Sept.)		59·31	5·94		58·95	13·52		57·18	11·46
Autumn (Oct. to Dec.)		46·77	10·61		46·56	5·53		40·76	3·25

NOTES. 1827.—The coldest *February* but one in the cycle: compare 1830, 1836, 1838, 1841; and *March* among the wettest: compare 1836, 1839. *April* the driest but one: compare 1839. *June* the driest but one: compare 1826. *July* among the warmest: compare back and forward. *September* very warm and dry. *October*, of the warmest and wettest: compare the rest. This is the middle year of five in succession having the barometrical average *above the mean of the cycle*: compare 1837 in Note.

1828.—The wettest *year* but one, and of the two warmest: compare 1826, also 1834: the *July* the wettest *month* in the cycle: compare 1827, 1834. Note, also, the gradation from year to year in the mean temperature of summer (first rising and then falling) through the space from 1824 to 1829. In the spring temperature, with a single exception, the like rise and fall. *April* was among the wettest: compare 1825. See figs. 5 and 6.

1829.—A cold year, and of the driest. *January* among the coldest of the cycle; and of the driest. The driest *March* but one: compare 1840: and among the coldest: compare 1833, 1836, 1837, 1839, 1840. *May* the driest of the series: compare 1826, 1833, 1834, 1836, 1839. *August* the wettest of the series: compare 1825, 1832, 1833. *September* wet. *December* the coldest in the set, and of the driest: compare 1830, 1835, 1839, 1840.

Mean height of the Barometer, Mean Temperature and depth
of Rain in each month, at Ackworth, Yorkshire, through a
Cycle of Eighteen Years.

	1830.			1831.			1832.		
	in.	°	in.	in.	°	in.	in.	°	in.
January	29·98	32·00	0·71	29·84	34·58	2·56	29·92	37·11	1·21
February.................	29·81	36·32	3·12	29·75	40·43	1·08	30·06	38·12	0·24
March	29·97	46·24	0·42	29·81	44·63	1·62	29·81	42·98	1·17
April	29·64	48·50	3·44	29·70	48·27	1·60	30·01	46·35	2·53
May.......................	29·80	51·10	3·75	29·92	51·53	2·18	29·92	51·05	1·92
June......................	29·74	55·00	4·88	29·87	58·42	2·11	29·82	58·70	4·09
July.......................	29·84	61·55	4·37	29·92	61·66	4·20	30·03	59·34	1·57
August...................	29·78	56·90	2·36	29·90	62·29	2·54	29·80	59·11	4·66
September	29·65	53·68	3·79	29·87	56·22	2·40	30·05	56·18	0·69
October	30·15	51·22	0·32	29·74	53·93	2·58	29·94	51·22	1·47
November	29·72	43·88	2·21	29·79	40·53	3·60	29·99	40·17	2·78
December	29·62	34·56	2·21	29·72	42·33	1·90	29·86	40·94	2·61
Whole year..............	29·81	47·58	31·58	29·82	49·57	28·37	29·93	48·44	24·94
Winter (Jan. to Mar.)...		38·19	4·25		39·88	5·26		39·40	2·62
Spring (Apr. to June)...		51·53	11·07		52·74	5·89		52·03	8·54
Summer (July to Sept.)		57·38	10·52		60·06	9·14		58·21	6·92
Autumn (Oct. to Dec.)		43·22	4·74		45·59	8·08		44·11	6·86

NOTES. 1830.—This year stands third in wetness, and shows
a low temperature. *January* and *March* are again very dry, and
the *February* between of the wettest! *April* to *July*, and *September*, afterwards, all show an excess of rain. The curious
equality of rain, in the winter and autumn moderate, and in the
other two seasons as excessive, is also worth notice. The temperature is again low, and that of *January* and *March* frosty.

1831.—The rain now falls off again, while the temperature
rises. The *July* and *November* are wet, the latter in the extreme
for this side of the cycle. The month of *August* presents a
great advance in warmth upon the preceding years, and the
whole year has a temperature above the mean.

1832.—This year closes the warmer series of the cycle. With
a still further approach to dryness, reducing the rain to a quantity below the average, we have also a reduction of the mean
temperature. The winter is of the driest, exceeded only by 1825,
and *February* has the least wet of any month but one in the
cycle. The month of *June* is of the wettest, and the rain of
August exceeded only by the same month in 1829.

Mean height of the Barometer, Mean Temperature and depth of Rain in each month, at Ackworth, Yorkshire, through a Cycle of Eighteen Years.

	1833.			1834.			1835.		
	in.	°	in.	in.	°	in.	in.	°	in.
January	30·21	33·66	0·76	29·72	43·54	3·70	29·98	34·52	1·77
February	29·41	41·70	2·76	30·02	42·06	0·57	29·60	40·79	2·69
March	29·88	38·45	1·34	30·11	44·25	1·49	29·85	41·22	1·67
April	29·72	46·07	2·78	30·15	45·17	1·83	30·06	45·20	1·06
May	30·05	57·93	0·54	29·99	53·80	0·66	29·81	50·35	2·60
June	29·70	58·03	3·11	29·89	58·77	1·99	29·97	57·15	2·28
July	29·94	60·42	1·09	29·91	62·01	7·03	29·96	60·53	0·63
August	29·87	56·30	4·50	29·83	60·20	2·09	29·92	61·52	1·55
September	29·86	53·03	1·49	30·04	55·82	1·83	29·60	53·95	2·33
October	29·74	48·73	2·76	29·93	48·92	0·26	29·66	45·18	2·47
November	29·80	42·93	1·32	29·92	45·05	1·16	29·84	41·70	2·02
December	29·53	43·11	2·61	30·23	41·42	1·13	30·11	35·68	0·12
Whole year	29·81	48·36	25·06	29·98	50·08	23·74	29·86	47·31	21·19
Winter (Jan. to Mar.)...		37·94	4·86		43·28	5·76		38·84	6·13
Spring (Apr. to June)...		54·01	6·43		52·58	4·48		50·90	5·94
Summer (July to Sept.)		56·75	7·08		59·34	10·95		58·67	4·51
Autumn (Oct. to Dec.)		44·92	6·69		45·13	2·55		40·85	4·61

NOTES. 1833.—This is a year of mean temperature, and having an average amount of rain. *January* of the driest and coldest: compare 1824, 1825, 1826, 1829, 1830, 1838. *February* wet: compare 1830, 1835, 1836, 1837, 1838. *March* dry and cold, and *April* wet. *May* of the driest, and the *warmest* of the series, but *variable*: thermometer at 80° and at 34°. *June* wet, *July* of the driest, and *August* wet: compare with the rest.

1834.—A warm year, in the *cold* period of nine; as 1829 was cold amidst the *warm* years. It has the highest barometer, and continues the tendency toward the *dry* extreme; its rains fell chiefly in *January*—(the wettest of the series, compare 1831 and 1837) and in July, the wettest, save in 1828, and *warm*. The *May* was dry, like the last, and the *October* the driest of the series.

1835.—A *year* of extreme *dryness*, second only to 1826, but the reverse of that in temperature, being of the coldest. *January* and *December* cold; and the latter, the driest *month* in the *whole cycle* of eighteen years: at this point the gradual tendency to drought ceases, and the *three* following years have each a mean amount of rain. The rain of this year falling so much in May and June, with July and August dry and warm, it may be thought to have been favourable to the crops, the general drought notwithstanding.

Mean height of the Barometer, Mean Temperature and depth of Rain in each month, at Ackworth, Yorkshire, through a Cycle of Eighteen Years.

	1836.			1837.			1838.		
	in.	°	in.	in.	°	in.	in.	°	in.
January	29·74	36·48	1·65	29·81	36·54	3·28	29·86	29·45	1·00
February	29·66	36·19	2·02	29·76	41·73	3·16	29·55	30·25	3·35
March	29·38	39·75	3·39	29·88	35·79	0·99	29·70	41·39	1·40
April	29·75	41·90	1·48	29·51	39·65	2·00	29·67	42·78	1·58
May	30·17	49·20	0·68	29·82	48·08	1·62	29·84	48·17	3·38
June......................	29·76	58·86	2·72	29·90	56·96	1·51	29·76	57·54	2·61
July	29·81	59·05	2·33	29·84	60·58	2·64	29·85	60·25	1·82
August....................	29·93	56·87	0·69	29·82	57·60	1·39	29·73	59·24	2·59
September	29·75	53·15	2·10	29·82	53·03	2·13	29·92	53·00	1·41
October	29·70	47·50	1·71	29·94	49·29	2·08	29·85	50·10	3·31
November	29·45	39·80	4·67	29·68	40·88	1·65	29·47	39·20	2·01
December	29·66	39·54	1·77	29·82	39·74	2·94	29·95	38·08	0·56
Whole year.............	29·73	46·52	25·21	29·80	46·66	25·39	29·76	45·79	25·02
Winter (Jan. to Mar.)...		37·44	7·06		38·02	7·43		33·69	5·75
Spring (Apr. to June)...		49·99	4·88		48·23	5·13		49·49	7·57
Summer (July to Sept.)		59·69	5·12		57·07	6·16		57·49	5·82
Autumn (Oct. to Dec.)		42·28	8·15		43·30	6·67		40·85	5·88

NOTES. 1836.—A *cold* year, the rain an average: a wet *February* and very wet *March*; with *April* and *May* dry: compare 1826, 1827, 1830, 1833, 1837, 1838, &c. The low temperature is pretty uniform through the twelve months. The *August* has the least rain of any month in the cycle of that denomination, and *November* the most: compare 1826, 1831, 1839, 1840.

1837.—The mean temperature and rain nearly as in last year. *January* and *February* in the wet extreme; *March* of the driest, and *April* reasonably wet. *December* the wettest of any in the cold years. The middle year of five *having the barometrical average below the mean of the cycle.*

1838.—The coldest *year* of the whole cycle of eighteen, chiefly due to the low temperature of the winter months: the third year in succession with rain about the average. *December* as dry as the preceding one was wet: *February, May* and *October* show the largest amount of rain. By *rain* the reader must, of course, sometimes understand the product of the gauge in *melted snow*: and the lying of this over, unmelted, may sometimes have thrown some excess upon the measure of the month following.

Mean height of the Barometer, Mean Temperature and depth of Rain in each month, at Ackworth, Yorkshire, through a Cycle of Eighteen Years.

	1839.			1840.			1841.		
	in.	°	in.	in.	°	in.	in.	°	in.
January	29·92	36·82	1·21	29·62	36·99	1·92	29·74	33·47	1·92
February	29·46	39·35	1·94	29·86	38·38	1·89	29·74	36·50	1·03
March	29·73	38·70	3·20	30·23	39·82	0·11	29·75	46·70	0·89
April	30·09	44·27	0·52	30·05	49·85	1·51	29·76	46·00	0·98
May	29·95	50·00	0·51	29·83	51·35	3·67	29·76	54·28	3·22
June	29·82	56·00	4·72	29·88	57·26	1·99	29·83	54·40	3·45
July	29·76	58·00	5·23	29·74	57·00	2·71	29·71	57·00	4·19
August	29·91	57·47	2·57	29·79	60·50	2·04	29·75	61·35	2·62
September	29·53	54·00	3·40	29·64	51·50	2·87	29·69	56·28	4·11
October	30·00	47·70	3·54	29·90	45·65	0·92	29·48	47·50	4·24
November	29·63	42·00	4·50	29·50	41·50	4·61	29·63	39·94	2·85
December	29·64	36·00	1·82	30·09	34·48	0·51	29·53	39·43	1·33
Whole year	29·78	46·69	33·16	29·84	47·02	24·75	29·70	47·74	30·81
Winter (Jan. to Mar.)...		38·29	6·35		38·39	3·92		38·89	3·84
Spring (Apr. to June)...		50·09	5·75		52·82	7·17		51·56	7·65
Summer (July to Sept.)		56·49	11·20		56·33	7·62		58·21	10·92
Autumn (Oct. to Dec.)		41·90	9·86		40·54	6·04		42·29	8·42

NOTES. 1839.—The wettest *year* of the whole cycle, with the lowest barometer, and also cold in the extreme; but the average temperature has risen nearly a degree. If we compare the rain assigned to the "Summer," with that of the same season in the *warm* 1828, and the *cold* 1829, we shall see that the wetness of a summer is not dependent on the temperature, but springs from other causes. The *April* and *May* of this wet and cold year are the poorest in rain of any of these (taken together) in the whole cycle.

1840.—The mean temperature still a little better, the rain considerably below average: *March*, showing only 0·11 in., should seem the driest *month* of the whole *cycle*, but that *snow*, fallen in the latter days of it, may have caused some part of the measure to go to the account of *April*; which is sufficiently meagre of rain in itself. The wettest *May* but one, and the like of *November*: compare 1830, 1836.

1841.—The mean temperature still advancing; but still exceeded by all the warm-period years, 1829 excepted: the rain up again to the wet extreme. *March* was pretty dry, as were the colder months generally: the rain, as to its excess, belongs to the warmer six months from *May* to *October*. The temperature of *July* in this year falls below that of the same month in every year of the warm period, save 1832: and *September* and *October*, taken together, are the wettest since 1824. Thus we complete the *cycle*.

PRINTED BY RICHARD AND JOHN E. TAYLOR,

RED LION COURT, FLEET STREET.

CORRIGENDA IN "THE CYCLE OF THE SEASONS, &c."

A *single figure*, taken wrong in copying a leading table, has given rise to the following; which oversight, the Reader is desired to excuse; more especially as the Cycle is improved in symmetry by the corrections. L. H.

Page 6, line 26, 48.152 read 48.126—48.952 read 48.879

27, 47.352 read 47.374

28, 1.6 read 1.405

11, line 5, 49.19 read 48.53

14, 48.952 read 48.879 ; and 47.352 read 47.374

15, 48.152 read 48.126 ; and 1.60 read 1.405

18, 49.241 read 49.159 ; and 2.230 read 2.148

12, line 3 from bottom, 39.856 read 39.413

2 ditto 48.105 read 48.069

16, in the Table, opposite " December" 48.54 read 40.54

" Whole year" 49.19 read 48.53

" Autumn" 46.93 read 44.26

And *dele* the words " mildest and the" in the *note*.

Errata, page 10, line 4 from bottom 0.42 read 042

12 ditto 0.44 read .044

Fig. 2.

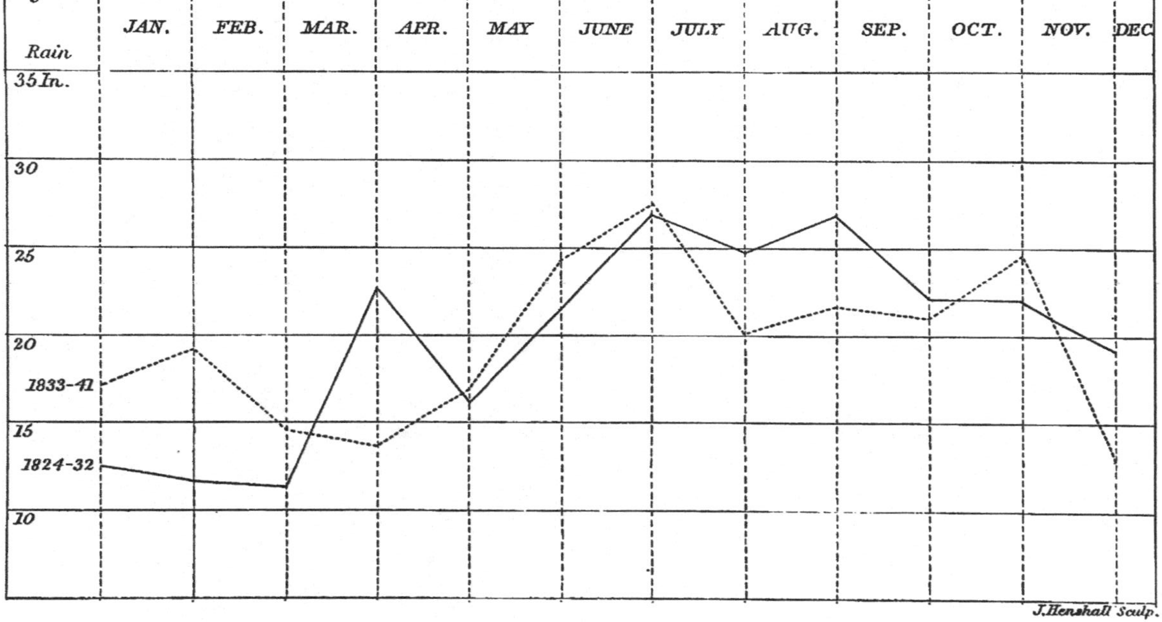

Rain
35 In.

30

25

20 — 1833-41
15
1824-32

10

JAN. FEB. MAR. APR. MAY JUNE JULY AUG. SEP. OCT. NOV. DEC.

J. Henshall sculp.

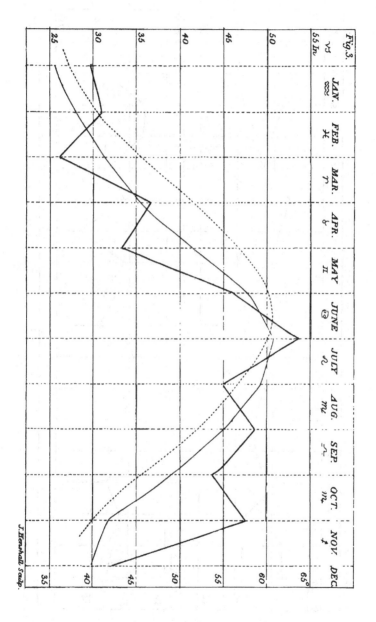

Fig.3.

| | JAN. | FEB. | MAR. | APR. | MAY | JUNE | JULY | AUG. | SEP. | OCT. | NOV. | DEC. |
| | ♒ | ♓ | ♈ | ♉ | ♊ | ♋ | ♌ | ♏ | ♎ | ♏ | ♐ | |

J.Kenshall Sculp.

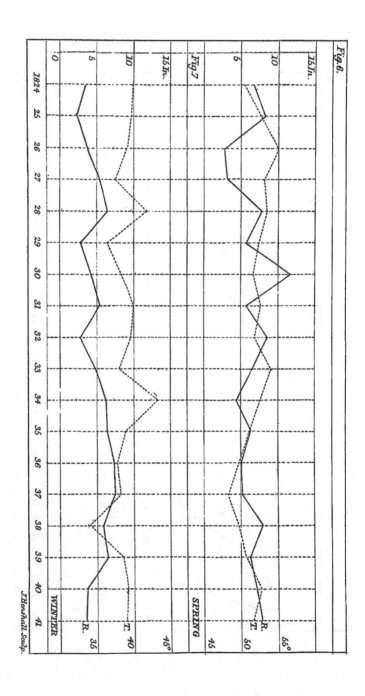

Fig. 6.

Fig. 7.

J. Henshall Sculp.

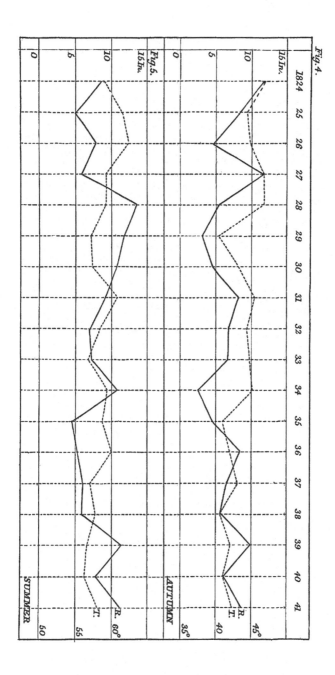

Fig. 4.

Printed in the United States
By Bookmasters